Active Learning

ACS SYMPOSIUM SERIES **970**

Active Learning

Models from the Analytical Sciences

Patricia Ann Mabrouk, Editor
Northeastern University

Sponsored by the
ACS Divisions of Analytical Chemistry and Chemical Education

American Chemical Society, Washington, DC

Library of Congress Cataloging-in-Publication Data

Active learning : models from the analytical sciences / Patricia Ann Mabrouk, editor ; sponsored by the ACS Divisions of Analytical Chemistry and Chemical Education.

p. cm.—(ACS symposium series ; 970)

Includes bibliographical references and index.

ISBN 978–0–8412–7404–4 (alk. paper)

1. Chemistry—Study and teaching—Congresses. 2. Active learning—Congresses.

I. Mabrouk, Patricia Ann. II. American Chemical Society. Division of Analytical Chemistry. III. American Chemical Society. Division of Chemical Education.

QD40.A28 2007
543.071—dc22

2007060809

The paper used in this publication meets the minimum requirements of American National Standard for Information Sciences—Permanence of Paper for Printed Library Materials, ANSI Z39.48–1984.

Distributed by Oxford University Press

PRINTED IN THE UNITED STATES OF AMERICA

Foreword

The ACS Symposium Series was first published in 1974 to provide a mechanism for publishing symposia quickly in book form. The purpose of the series is to publish timely, comprehensive books developed from ACS sponsored symposia based on current scientific research. Occasionally, books are developed from symposia sponsored by other organizations when the topic is of keen interest to the chemistry audience.

Before agreeing to publish a book, the proposed table of contents is reviewed for appropriate and comprehensive coverage and for interest to the audience. Some papers may be excluded to better focus the book; others may be added to provide comprehensiveness. When appropriate, overview or introductory chapters are added. Drafts of chapters are peer-reviewed prior to final acceptance or rejection, and manuscripts are prepared in camera-ready format.

As a rule, only original research papers and original review papers are included in the volumes. Verbatim reproductions of previously published papers are not accepted.

ACS Books Department

Contents

Problem-Indexes

Preface

About 10 years ago, supported by the National Science Foundation, nearly 60 leading analytical researchers, educators, and administrators from academe, government, and the private sector met to discuss the marked changes occurring in the world relative to the practice of analytical chemistry and to discuss the requisite changes in the education and training of analytical scientists. The resulting report *Curricular Developments in the Analytical Sciences* authored by Professor Theodore Kuwana (University of Kansas) called for marked changes not only in course content but also in the mode of delivery. The report offered a number of formal recommendations punctuated by a call for the investigation of active-learning methods and problem-based learning, in particular. This fledgling, grassroots effort burgeoned during the past decade into a thriving discipline-wide reform movement punctuated by a number of long-standing symposia and workshops at the Eastern Analytical Symposium, the Pittsburgh Conference, American Chemical Society (ACS) National Meetings, a notable series of "A-page" articles in *Analytical Chemistry*, as well as other notable activities and projects including the Analytical Sciences Digital Library, some of which are discussed in this volume.

This ACS Symposium series volume is partially based on a symposium *Innovative Approaches for Teaching Analytical Chemistry* organized and chaired by Patricia Mabrouk, sponsored by the ACS Divisions of Analytical Chemistry and Chemical Education, which took place at the 230th National Meeting of the ACS in Washington, D.C., August 28–September 1, 2005. Many of the contributors to this volume were speakers in this symposium or at the J. Calvin Giddings Award Symposium that honored Professor Frank Settle (Washington and Lee College), a leader in the reform movement. The Award Symposium occurred concurrently with the ACS National meeting.

This book strives to present a balanced picture of how the education and training of analytical chemists has changed in the decade since the NSF-sponsored curricular workshops. These workshops spawned a wide array of active learning activities at the college–university level. This volume provides a number of examples of how these methods have been applied in teaching analytical science, broadly defined, at colleges and universities at both the undergraduate and graduate levels. Contributing authors also discuss how our discipline and academe in general must continue to evolve and change if our nation is to continue its leadership in science, technology, engineering, and mathematics.

This volume is a first in many ways—several books have been published in recent years that focus specifically on problem-based learning or other specific active learning methods but none have neither discussed active learning methods broadly per se nor illustrated their use in teaching a specific discipline. As such, the strategies, materials, and experiments described in the volume should be of interest to practicing analytical chemists, chemical educators, and chemistry teaching faculty and graduate students no matter their area of specialization. It is hoped that this volume will serve as a catalyst to promote further discussion, thoughtful experimentation in the classroom and the laboratory, innovation, and reform not only in the analytical sciences but also in the allied fields of science, technology, engineering, and mathematics.

In closing, the editor wishes to express her sincere thanks to each of the authors of the chapters of this volume. You are an amazingly talented, dedicated, and hardworking group. It has been a great pleasure and honor to partner with you. I wish each of you the very best and I look forward to seeing what the next decade brings as a result of your willingness to participate in the present "experiment!"

Patricia Ann Mabrouk
Department of Chemistry and Chemical Biology
Northeastern University
Boston, MA 02115

Chapter 1

Introduction

Patricia Ann Mabrouk

Department of Chemistry and Chemical Biology, Northeastern University, 360 Huntington Avenue, Boston, MA 02115

Active learning methods have been the subject of much interest in higher education since the turn of the twentieth century. However, during the last decade we have seen a dramatic increase in their application in the college classroom and laboratory. The current volume contains twenty chapters that focus on recent developments in active learning as related to the education of scientists in the analytical sciences. The purpose of this introductory chapter is to: (1) introduce active learning and its principal methods that will be discussed in this volume; (2) introduce the subjects to be discussed in the following chapters; and (3) highlight some important aspects of subjects not covered in this volume, that have future implications to the field.

Introduction

The past decade has been a period of significant change in higher education as institutions of higher learning have attempted to adapt to the rapid changes taking place in our world. This book is an effort to examine how education in the analytical sciences, broadly defined, at the undergraduate and graduate levels, has changed over the past decade in response to the marked intertwined technological, economic, and social changes occurring. These changes have tremendous significance not merely for our discipline but for the academy as a whole. Though academia have been employing more chemists recently, the majority of chemists find employment in the private sector (*1*). In her chapter,

Dr. Dorothy J. Phillips, Director of Strategic Marketing at Waters Corporation, sets the stage for the chapters that follow. She forcefully makes the case for active learning as it relates to the career of tomorrow's industrial chemist. They need to be trained with interdisciplinary skills including teamwork, problem solving, and transnational competency skills to assure our graduates can successfully meet the unknown challenges of tomorrow's global scientific workplace.

In order to fully appreciate where we are and where we are going in higher education in the analytical sciences, we need to first consider our roots – where we came from - how the field of analytical chemistry has matured as an academic discipline and a community of scholars. Consequently, Dr. Phillips' chapter is followed by a contribution from Professor Frank A. Settle (Washington and Lee University), providing a historical perspective on training in the analytical sciences in the United States, focused on the changes that have occurred over the past half-century. Since analytical chemistry is inherently the science of measurement, this chapter discusses how advances in electronics, computers, and analytical instrumentation as well as in our understanding of how we learn have impacted the analytical lecture and laboratory curriculum and supporting materials.

Many of the changes have been brought about by the information explosion resulting from the tremendous changes in digital technology. It has become virtually impossible to cover any subject in the college classroom in an authoritative way. This has forced each discipline to identify its core concepts and to focus on making sure that student learners master those concepts and teach students to be competent, self-directed, life-long learners. Active learning methods in particular have been emphasized by several groups (*2,3*) as particularly effective in this regard and have been adopted by a number of chemical educators for use in their classrooms. A number of the chapters in this book describe their classroom experiments with active learning in its various forms.

Active Learning

Active learning generally refers to those teaching techniques that actively engage students in the classroom. Active learning can be used with students working either independently or in small-groups, in small and large classes, at all levels. The term can be taken to mean any classroom activity that is more "active" than passive listening. Short-term activities that fall in this category include think-pair-share and the minute paper (*4*). In this book, however we will focus more on those methods that typically require students to work together collaboratively, utilizing multiple sensory modes (reading, writing, speaking, listening), and engaging in higher-order cognitive processes (analysis, synthesis, and evaluation) when interacting with new course content that may be presented in the form of open-ended, contextualized problems. Examples of these types of active learning methods include cooperative learning (CL), service learning, case study method, and problem-based learning (PBL).

Cooperative Learning

Cooperative learning is, in a nutshell, a form of active learning in which students work together in small groups to learn specific content. To truly qualify as being cooperative learning, five basic elements need to be present. These are: the establishment of a positive interdependence between student learners; individual accountability; meaningful, face-to-face interaction between students; social interaction; and group processing of information. Learning occurs in the collaborative process through which students exchange information. It can be either informal or formal. Groups can be constituted to work together either on a temporary basis on short term assignments lasting anywhere from minutes to a class period in duration, or for an extended period of time up to a semester on long-term projects. It is difficult to pinpoint when cooperative learning actually began. Many identify John Dewey as an important "Father" figure. In fact, the approach has enjoyed widespread use in elementary, middle, and high schools for decades. However, CL really came into use in the college and university classroom in the 1980's. In their contribution, Professor Emeritus Richard M. Felder (North Carolina State University) and Rebecca Brent (Education Designs, Inc.) provide an authoritative overview of the theory and practice of cooperative learning, together with a set of practical guidelines for those faculty interested in implementing quality CL experiences in their classrooms.

Problem-Based Learning and Project-Based Learning

Problem-based learning is a subset of CL in which students investigate open-ended and often ill-defined real-world problems. Learning results from group efforts to wrestle with these problems. PBL was originally developed and implemented at McMaster University in 1969 in the training of physicians. The method was quickly adopted by other medical schools. It came to the attention of many college and university science educators through the Boyer report (*2*) which offered the University of Delaware as an exemplar based on its success with problem-based learning. Today PBL enjoys international use across the curriculum and at all levels from K-16.

PBL resonates strongly within the analytical chemistry community. It was identified as a particularly promising teaching method in the 1996 report entitled "*Shaping the Future*" based on two workshops organized by Professor Emeritus Theodore Kuwana (University of Kansas) and supported by the National Science Foundation to discuss how to best equip tomorrow's analytical workforce (*3*). Consequently, this volume contains several chapters discussing various aspects of problem-based learning. A special variant of problem-based learning is project-based learning. This term usually describes the application of PBL in the laboratory. Professor Thomas J. Wenzel discusses his work in restructuring the undergraduate analytical chemistry curriculum at Bates College using a CL model in the classroom and project-based learning in the laboratory to promote deeper student learning. This is followed by a contribution from me that describes a unique take on PBL in which the laboratory project drives all of the course-related activities including the lectures in an advanced undergraduate chemistry elective with a focus on bioanalysis.

Case Study Method

Case studies, which provide a context-rich medium for active learning, are often used in cooperative learning, problem-based learning, and service learning. Case study typically involves the analysis of a brief story or "case" that has been thoughtfully designed to summarize the essential facts surrounding an event that illustrates some principle which the class is studying. Case study is different from project-based study in that no data collection is typically involved. The focus in case study is on the analysis of an event that took place in the past though it may be useful in looking into the future.

Technically, it is difficult to pinpoint when case studies were first formally used in instruction. In the United States, the technique was first formally introduced over a century ago at the Harvard University Law School. Case study teaching was first introduced in the field of chemistry by Professor James B. Conant in the early 1940s upon his return to Harvard University following work on the Manhattan Project (*5*). Today the method enjoys widespread use at all levels and in all fields of higher education. In their contribution to this volume, Professors Simon T. Belt and Tina L. Overton discuss the use of context-based (pharmaceutical, environmental and industrial chemistry) case studies in teaching undergraduate analytical chemistry at several universities in the United Kingdom. The authors present examples of the use of case studies both in the traditional classroom environment and on-line through a virtual learning environment.

Service-Learning

Service-learning (SL) (*6,7*) is an experiential learning method which has been investigated by analytical chemists in their undergraduate classrooms in recent years. In service learning, students partner with their local communities to provide information and/or services that benefit both the provider and the recipients. The connection between the classroom and the community, crafted thoughtfully and intentionally, enriches the student's learning experience in unique ways developing not only the student's knowledge content and higher cognitive skills but teaching effective communications skills, civic pride and responsibility, ethics, public policy, and racial and cross cultural understanding and sensitivity. Professor Alanah Fitch (Loyola University Chicago) reviews the history of service-learning in chemical education over the past decade, as reflected by publications on service-learning, that have appeared in the peer-reviewed literature. In the following chapter, Professor Anna G. Cavinato provides two examples of service-learning projects with an environmental focus in use at Eastern Oregon University. The first project serves as a vehicle to introduce key analytical chemistry concepts and instrumentation in the General Chemistry course. The second example illustrates the application of the service learning model in an upper-level environmental chemistry course.

Online or Electronic Methods

Given the pervasiveness of computer technology in our society, it is not surprising that today's college students are heavy users of various types of digital technology including the internet, email, and instant messaging (*8*). The majority of college students have their own computers and view the internet as a standard information retrieval tool. In fact, they are more likely (73%) to use the internet than their college or university library in order to locate information. Although it is really too early to state with any certainty, some evidence suggests that the internet and computer-based instruction offer a number of advantages including anonymity, social equality (gender and race), and convenience. Many faculty are experimenting in their classroom with emerging forms of computer-based technology such pod casts (*9*), blogs (*10*), and wikis (*11,12*). The emergent digital technology has tremendous educational potential to afford students self-paced, interactive, personalized and private instruction. Many forms (but not all) present potential tools in support of active learning as they have the potential to place the student squarely in control of the information retrieval and problem-solving process (a "must" in active learning). For example, animations, both passive and active, offer the opportunity to help students construct robust cognitive models of sophisticated, yet fundamental, scientific concepts such as the particulate nature of matter. Simulations provide students equal access to sophisticated, expensive analytical instrumentation that might ordinarily be inaccessible to the student due to cost or availability. These resources allow students to conduct potentially dangerous or costly, yet pedagogically valuable, experiments (*e.g.*, involving radioactivity or biohazards) safely. They also provide students, when needed, the ability to repeat these experiments maximizing their learning potential. Finally, these resources potentially provide on-time delivery of instruction facilitating the critical timely exchange of needed information between students and caring educators.

In view of the enormous potential of computer-based technology to facilitate training in the analytical sciences, it should not be surprising that a significant number of faculty in the analytical sciences have already investigated its potential to impact student learning in their classrooms. A number of the chapters in this volume explore different applications of digital technology including animations, interactive software, digital libraries, and online learning and how they can be used in support of active learning.

Animations

As mentioned above, computer animations may improve student understanding of chemistry by improving their ability to visualize key chemical concepts such as the particulate nature of matter and dynamic chemical processes (*13*). In his chapter, Professor Thomas G. Chasteen (Sam Houston State University) reviews the application of both static and interactive computer animations throughout the analytical chemistry curriculum, and provides a number of nice examples of both static/passive use and interactive animations that he and others have used to teach a wide range of topics such as chromatography and spectroscopy. His chapter discusses the benefits of interactive animations in promoting deeper understanding of the increasingly complex technical information with which analytical scientists must wrestle and introduces some of the currently available software which educators can use to generate their own interactive digital media. Finally, this chapter identifies a number of important issues that faculty interested in developing and posting animations should consider at the outset of their development effort.

Software Applications

As Professor John H. Kalivas (Idaho State University) describes in his chapter, application of active learning methods and recent developments in digital technology, have made it possible to introduce some fairly sophisticated and powerful mathematical tools which analytical chemists use in research today such as factor analysis and principal component analysis. They can be applied in a meaningful way in the undergraduate chemistry curriculum. Drs. Marie Scandone, Deborah Kernan and Gregory M. Banik from Bio-Rad Laboratories, Inc. review a number of the commercially available software including their own KnowItAll® suite. It is available at no cost for use by academia. The software can be used to draw and visualize complex chemical structures or to identify chemical structures through mass spectral, infrared, NMR, and/or Raman spectroscopy. Professors Mark E. Bier (Carnegie Mellon University) and Joseph J. Grabowski (University of Pittsburgh) provide a powerful example of web-based interactive instruction using the case study method in their chapter introducing the "Virtual Mass Spectrometry Laboratory" (VMSL). VMSL provides undergraduates experience in data collection and analysis via instrument simulations with state-of-the-art mass spectrometers. At the time of this publication, mass spectrometers are quite expensive, somewhat temperamental and unavailable for "hands on" student use at most colleges and universities. The electronic medium allows students to explore the application of mass spectrometry in many areas such as proteomics, polymer analysis, forensics, and small molecule analysis.

Digital Libraries

Recognizing the growing body of high quality, electronically accessible educational materials and the need to develop an inventory of these materials for use by educators and students at all educational levels across the STEM disciplines, the National Science Digital Library (NSDL) program (http://nsdl.org/index.php) was conceived within the National Science Foundation's (NSF) Division of Undergraduate Education (DUE) in 1995 (*14*). The first digital library prototypes were funded in 1998 and the first formal round of funding occurred in 2000. Nearly 200 NSDL projects have been funded to date. In their chapter, Professor Cindy K. Larive (University of California, Riverside) and Professor Emeritus Theodore Kuwana (University of Kansas) report on a relatively new NSDL project in the analytical sciences, the Analytical Sciences Digital Library (ASDL; avail. URL: www.asdlib.org). This project uses peer-review to identify and catalogue reliable, robust web-based resources of value to both faculty and students in the analytical sciences on topics ranging from active learning techniques to resources on specific technical content related to analytical chemistry.

Distance Learning

Digital technology has transformed, even revolutionized, distance learning which, of course, can benefit strongly from the aforementioned advances in interactive software, animations, discussion boards, wikis, etc. Preliminary evidence suggests that distance learning may be very effective in promoting deeper student learning and greater satisfaction with the learning experience. The most recent National Survey of Student Engagement observed that distance learning students self-reported greater educational gains, were more frequently engaged in deep learning activities including reflective learning activities, and experienced a higher level of satisfaction with their college experience than their campus-based peers (*15*). In his contribution, Professor Roger K. Gilpin and C. S. Gilpin discuss key issues in the development and implementation of quality online courses. Prof. Gilpin offers as an example his own experiences in developing and implementing a one-semester introductory chemistry course at Wright State University focused on fundamental analytical principles. The course uses a variety of interactive computer-based activities, animations, and Excel®-based simulations to promote deeper student learning.

Training Chemical Technicians

Many students receive their academic training in our nation's community college system. In fact community college students constitute 45% of all students enrolled in U.S. colleges and universities (*16*). Community colleges represent the primary source of technician education for our nation (*17*). Professor John Kenkel (Southeast Community College) reviews the history of the development of the "ChemTechStandards" and standards-compliant educational materials for the training of chemical technicians in the analytical sciences at two-year institutions, and provides twelve examples of the analytical chemistry curriculum and assessment techniques at ACS certified two-year chemistry laboratory technology programs.

Course Design and Assessment

At the undergraduate level, there are two standard course offerings in analytical chemistry: Quantitative Analysis (QA) and Instrumental Analysis.

Professor Robert J. Eierman reviews the history of QA in terms of the impact of active learning methods on student learning and presents a redesign and assessment project, based on backward design, called the "New Quant Project," which has been structured to enable meaningful conclusions to be drawn concerning the relationship between the educational process and outcomes.

Graduate Education

Critical concerns have been raised recently about graduate education (*18*) including the importance of properly orienting new students early within their university program in order to retain them (*19,20*), the need to foster creative thinking (*21*), provide educational depth and yet breadth (*21*), and for better career counseling. Professors Alexander Scheeline and Ryan C. Bailey examine the value of active learning in graduate training in developing student self-direction, information literacy, and the higher-level cognitive skills (comprehension, application, analysis, synthesis, and evaluation). The authors illustrate the potential benefits for a self-described "boot camp" first-semester graduate survey course on analytical chemistry required of all entering analytical chemistry graduate students at the University of Illinois at Urbana-Champaign.

Changes in analytical instrumentation and experimental methods have led to marked changes in the complexity and fundamental nature of the problems that scientists can tackle. Increasingly, we are seeing evidence that a collaborative model is needed to address today's research problems. The increasingly global research environment often necessitates collaborative work between scientists in different disciplines and from different cultures (*22*). Consequently, it is critical that our students possess strong team work and communication skills. These issues are explored with a focus on graduate training by Professors William R. Heineman (University of Cincinnati) and Richard N. Zare (Stanford University) in the final chapter in this volume. Active learning, by its very nature, promotes the development of a wide range of cognitive and affective skills in our students as they partner together with other highly trained professionals.

Future Perspectives

Until recently, the majority of students pursuing degrees in STEM in the U.S. were white males. Today more than half of all undergraduates are women, yet women still account for less than 20% of all undergraduate degrees granted in the STEM disciplines (*23*). Furthermore, the majority of these women pursue degrees in biology, psychology, etc. Women and minorities continue to be underrepresented for higher degrees. The situation is bleaker for Blacks and Hispanics. Some of the key factors contributing to the paucity of underrepresented minorities in STEM disciplines are known to include poor student academic preparation, a perceived negative social image of scientists, and a lack of encouragement (*24*). This means that a significant fraction of our nation's potential STEM talent pool is literally being thrown away. This is not a subject directly addressed in this volume but one we must address if we are to be successful. There is a growing body of evidence suggesting that active learning methods, applications of digital technology, etc. may positively impact the diversity of the STEM talent pool (*25*). However, much more than this will be required in order to attract and more importantly retain these students in STEM careers (role models, peer mentoring, financial aid, etc.).

Assessment

Several authors, including Bier, Grabowski, and Eierman have noted the need to think about effective methods for assessing educational experiments. Rigor in assessment has really only come to the forefront over the last decade. To scientists, this is somewhat of a "no brainer" as our students would say. Certainly, we all recognize that there is no way to know if educational experiments are effective unless one assesses systematically using rigorous methods. That said, the issue of assessment is becoming somewhat divisive within the STEM educational community. Many STEM faculty do not possess formal training in education or psychology and consequently do not know enough about assessment techniques to accomplish summative assessment of their classroom experiments in a meaningful way. This point was brought home to me while I was putting this volume together. We have reached a crossroads in science education. Increasingly, thoughtful, caring educators are being asked to be experts in their technical fields, education, and assessment. The standards imposed in reporting classroom and laboratory experiments are becoming quite high. In the long run, this will lead to a strong knowledge base about effective teaching, learning and assessment methods. However, in the short run, it may prohibit some sharing of interesting ideas with the greater science education community. This would be a tragedy for all: students in their classrooms and the community.

I think the answer is clear. We have heard a strong call for collaboration from Professors Heineman and Zare in this volume. I believe we need to form educational partnerships between analytical scientists possessing technical expertise and science educators, cognitive psychologists, assessment professionals and experts from other fields both from within the academy and in society. This will enable the teaching and learning of increasingly sophisticated analytical techniques using educational methods that have been thoughtfully and rigorously assessed and shown to be efficient and effective pedagogically. However, this will require significant changes in higher education and call for educational institutions and their administrations to value and support in meaningful ways (core values – tenure, promotion, merit review) such collaborative educational efforts.

Academic Partnerships with Community Colleges

As Professor John Kenkel has illustrated in his chapter, much fine and thoughtful curricular work in the analytical sciences is taking place at two-year colleges. Little of this work currently appears in the peer-reviewed literature. This is likely due to the high teaching loads of our nation's community college educators. Since community colleges now enroll nearly half of all undergraduates (*26*) many of whom will transfer to four-year institutions (44 % of all science and engineering graduates attend community college at some point in their education (*17*)), it is critical for all educational stakeholders to be at the table and actively participate in decision making processes that affect their classrooms and ours (nearly 60% of undergraduates today attend more than one academic institution while pursuing their degree (*26*)). Work needs to be done to better knit the educational community of two-year college science educators with their counterparts at four-year institutions to share ideas and resources and establish and pursue collaborative (yes, again!) efforts addressing the unique needs of today's students.

Final Thoughts

Active Learning: Models from the Analytical Sciences does not focus on specific course or discipline-specific content. The goal is not to identify or specify the technical content that analytical chemistry faculty should teach in their classrooms. Rather, the volume is intended to provide readers, who may or may not be analytical chemistry faculty, with a practical handbook of information about promising active learning teaching practices and technologies. As such, the volume offers the reader an overview of current and emergent teaching practices and tested guidelines for experimenting with and changing the ways they teach. The volume does not make any assumptions regarding the readers' knowledge of or experience with learning theories, teaching methodologies, or analytical chemistry. Each chapter has been written so the reader can read and digest it in whatever order he or she prefers.

I hope that this volume challenges you to regularly read the chemistry and interdisciplinary science education journals such as *Journal of Chemical Education*, *The Chemical Educator*, and *Journal of College Science Teaching*, to attend and participate in local, regional and national science education conferences, and professional, or electronic discussion groups, and to actively participate in the vital ongoing active learning experiment taking place inside today's college classroom. I refer the interested reader to the sources listed in the reference section in each chapter as a starting point for further study. For those readers who wish more scholarly depth, I have prepared a brief listing of valuable archival resources. This list appears at the end of this volume together with some suggestions for those faculty interested in pursuing their own classroom and/or teaching laboratory experiments.

References

1. Heylin, M. *Chem. Eng. News* **2005,** *83*, 41-51.
2. The Boyer Commission on Educating Undergraduates in the Research University. "Reinventing Undergraduate Education: A Blueprint for America's Research Universities," Carnegie Foundation for the Advancement of Teaching, 1998.
3. Kuwana, T. "Curricular Developments in the Analytical Sciences," National Science Foundation, 1998.
4. McKeachie, W. J.; Svinicki, M. *McKeachie's Teaching Tips: Strategies, Research And Theory for College And University Teachers*; 12 ed.; Houghton Mifflin Company: Boston, 2005.
5. Conant, J. B. *The Growth of the Experimental Sciences: An Experiment in General Education*; Yale University Press: New Haven, 1949.
6. *Service-Learning in Higher Education: Concepts and Practices*; Jacoby, B., Ed.; Jossey-Bass, Inc.: San Francisco, 1996.
7. The National Service-Learning Clearinghouse. Service Learning. http://www.servicelearning.org/index.php (accessed December 9, 2006).
8. Jones, S. "The Internet Goes to College," Pew Internet and American Life Project, 2002.
9. Read, B. *Chronicle Higher Educ.* **2005**, *52*, A39.
10. Downes, S. *Educause Rev.* **2004**, *39*, 14-26.
11. Lamb, B. *Educause Rev.* **2004**, *39*, 36-48.
12. Read, B. *Chronicle Higher Educ.* **2006**, *53*, A31.
13. Yezierski, E. J.; Birk, J. P. *J. Chem. Educ.* **2006**, *83*, 954-960.
14. Zia, L. L. *D-Lib Magazine* **2001**, *7*, 1-8.

15. Education, I. U. C. f. P. "Engaged Learning: Fostering Success for All Students," Indiana University Center for Postsecondary Research, 2006.
16. American Association of Community Colleges. "Facts 2006," American Association of Community Colleges, 2006.
17. American Association of Community Colleges. "Teaching by Choice: Cultivating Exemplary Community College STEM Faculty," Community College Press, 2006.
18. *Envisioning the Future of Doctoral Education: Preparing Stewards of the Discipline - Carnegie Essays on the Doctorate* Golde, C. M.; Walker, G. E., Eds.; Jossey-Bass: San Francisco, 2006.
19. Lovitts, B. E. Doctoral thesis, University of Maryland, 1996.
20. Workman, M.; Bodner, G. *Chem. Educ.* **1996**, *1*, 1-12.
21. Jacobs, M. *Chem. Eng. News* **1996**, *74*, 40-43.
22. Burland, D. M.; Doyle, M. P.; Rogers, M. E.; Masciangioli, T. M. "Preparing Chemists and Chemical Engineers for a Globally Oriented Workforce: A Workshop Report to the Chemical Sciences Roundtable," National Academies Press, 2004.
23. National Science Foundation, Division of Science Resources Statistics. "Women, Minorities, and Persons with Disabilities in Science and Engineering: 2004," National Science Foundation, 2004.
24. "Land of Plenty: Diversity as America's Competitive Edge in Science, Engineering, and Technology," Congressional Commission on the Advancement of Women and Minorities in Science Engineering and Technology Development, 2000.
25. May, G. S.; Chubin, D. E. *J. Eng. Educ.* **2003**, *92*, 27-40.
26. Adelman, C. "Moving Into Town - and Moving On: The Community College in the Lives of Traditional-Age Students," U.S. Department of Education, 2005.

Chapter 2

An Industrial Perspective toward Analytical Chemical Education

Dorothy J. Phillips

Waters Corporation, 34 Maple Street, Milford, MA 01757

The active learning process is critical to a successful profession as an industrial chemist in the 21st Century. Active learning is a requirement not an option for analytical chemists in industry. The educational process that requires active not passive learning begins not ends when a scientist enters the industry sector. The knowledge that one brings to a new position is the basis on which the person grows. Since an analytical-industrial chemist must stay current with technology, learning is a nonstop process. This chapter covers the types of training that analytical chemists should bring to industry to be prepared for the chemical enterprise in 2015 and beyond. In addition, the chapter discusses the training analytical scientists should receive during their vocation to strengthen or apply their basic education.

Introduction

An extensive career in industry gives the scientist an opportunity to assess the importance of basic education and the success that is possible by participating in active learning throughout one's career. During an industrial career chemists have the opportunity to strengthen their multi-disciplinary academic training, for example, in biochemistry and analytical chemistry. Writing this chapter allows the author to give an industrial scientist's perspective on what are currently the valuable components in training analytical chemists for the future.

Analytical chemistry has been defined as the development and application of methods, instruments and strategies to obtain information on the composition and nature of matter in space and time (*1*). This field has been in great demand in industry because every chemical endeavor requires some sort of measurement to help answer a question, and analytical chemists should be experts at making valid measurements. Based on the demographics of the American Chemical Society (ACS) most analytical chemists work in industry About 60% of the more than 158,000 ACS members have chosen industrial over academic or government professions (*2*). In addition, the majority of the approximately 8% of ACS members who belong to the Division of Analytical Chemistry work in industry. The training required to be successful in industry depends on the positions analytical chemists should be prepared to fill in the 21st century.

Industries and Positions for Analytical Chemists

The 2005 President of ACS, Dr. William Carroll, led a Society-wide effort during his presidential succession period to develop scenarios of how the chemical enterprise (education, industry and government) would change in the next ten years (*3*). The goal of his program "Chemical Enterprise 2015" was to understand how ACS can help its current and future members adapt to these changes and how the society can take advantage of these new opportunities. The industry component of the chemical enterprise in the final report was extended beyond traditional chemicals and petrochemicals to include energy, biotechnology, and pharmaceuticals as well. The report indicates that the analytical chemist of 2015 is more likely to work in a small company instead of one of the large businesses with a readily recognizable name and the projects will cross the boundaries of science and often be multidisciplinary, forcing continuous training. The additional challenge will be outsourcing, moving jobs outside the boundaries of United States to Asia and India for cost-reduction and technology advantages. Traditional chemical companies and the pharmaceutical industry are investing in joint ventures for manufacturing in developing countries. Thus, the availability of jobs will decrease forcing companies to hire only the best and the brightest. The better prepared the analytical chemist the greater probability of getting an industrial position.

In agreement with Dr. Carroll's report, companies that require analytical chemists include (*1*) pharmaceutical firms (requiring expertise in mass spectrometry and separations instruments), and biomedical start-up companies (for example, biomedical analysis and instrumentation). Other industries requiring analytical chemists include the basic chemical industry, the energy industry (for example, petroleum and battery developers), electronics (for example, R&D in molecular diodes and LEDs) and also instrument manufacturers (for example, developers of liquid chromatography systems).

A 2005 issue of *Chemical and Engineering News* (C&EN) contained an article on positions held by analytical chemists today, "Analyze This, Analyze That" (*4*). Three analytical chemists interviewed in this article held very different positions; one scientist works in a traditional chemical company; another is a professor in a cell biology department; and the third is a Ph.D. chemist in analytical instrument sales. The projects for the chemist in a traditional chemical plant include optimizing the additive system in a plant, improving how products get to customers, and coordinating analytical services for a business unit. The professor's interdisciplinary research in cell biology includes mass spectrometry and proteomics. Traditional positions in universities did not cross boundaries of disciplines. In contrast analytical chemists in industry often work across departments or in interdisciplinary programs. The third analytical chemist has been selling analytical instruments for more than twenty-two years. At the time of the interview he was a mass spectrometry specialist interfacing with scientists in academic labs, pharmaceutical companies, police crime labs and other settings. The C&EN article also interviewed a scientific resource agency that arranges contract work for scientists; the firm places most analytical chemists in quality assurance/quality control spots in pharmaceutical, chemical, and food technology companies.

Many analytical chemists find themselves in pharmaceutical and medicine manufacturing companies (who are the principal employers of analytical chemists) that provided 291,000 wage and salary jobs in 2004 (*5*). Quality control and quality assurance are vital in the pharmaceutical industry. Many production workers are assigned on a full time basis to quality control and quality assurance functions, whereas other employees may devote part of their time to these functions. The caveat is that many of these positions are not available now and even less will be offered in ten years in the United States and Europe. These positions are among those that are being moved abroad to India and China.

Analytical chemists also hold positions at laboratory instruments manufacturers like Waters Corporation. These scientists analyze products in

R&D, applications development and manufacturing. After commercialization and throughout a product's lifecycle, it is supported by scientists in sales, technical support and service positions. Again these positions are not just in the major industrial countries; instrument manufacturers must also address the need to take manufacturing and its support overseas. For example, Waters announced in April 2006 the formation of a manufacturing partnership with the electronics manufacturer Solectron in Singapore (www.waters.com).

The question arises as to whether analytical education should shift its focus to prepare students for jobs in developing countries. Currently, many overseas jobs are being filled by Asian and Indian students trained in the United States and Europe. Their expertise in science combined with being multi-lingual makes them ideal candidates for these overseas positions.

This chapter will cover the chemical education that analytical chemists should receive to be ready for positions in the various industries. It will also cover the formats of continuous active learning.

What should the Undergraduate and Graduate Education include for the Analytical Chemists to be "industry ready"?

In line with ACS Past-President Carroll's initiative the theme for presidential events at ACS National meeting in Washington DC in 2005 was "Meeting the Workforce Needs of the Chemistry Enterprise in 2015". Are analytical chemists being educated today for an industrial career that spans to and beyond 2015? What is required during the academic period and what should be added once in industry? Is the chemist prepared to be a strong contributor in an industrial position after receiving a bachelor's degree, a masters or a doctoral degree? Does the ability to invent, market or sell new products increase after a post-doctoral period?

Formal training

The pharmaceutical industry practices offer a good example of the requirements for an analytical chemist to start a career and to grow in industry (5). Given the range of tasks that an analytical chemist can carry out in the pharmaceutical industry, having a solid, broad background in chemistry is highly desirable. More than 6 out of 10 of all workers have a bachelor's, master's, professional, or Ph.D. degree—more than twice the proportion for all industries combined. For science technician jobs in this industry, most companies prefer to hire graduates of technical institutes or junior colleges or those who have completed college courses in chemistry, biology, mathematics, or engineering.

Some companies, however, require science technicians to hold a bachelor's degree in a biological or chemical science. For higher level scientific and engineering jobs, a bachelors of science degree is the minimum requirement. Scientists involved in research and development usually have a master's or doctoral degree. In addition most sales representatives have at least a 4-year degree.

Analytical chemists can increase their marketability in the pharmaceutical industry by learning medicinal chemistry and biochemistry, and by gaining exposure to pharmacokinetics and chemoinformatics. Positions in many companies but especially the pharmaceutical industry require expertise in mass spectrometry (MS) and separations instrumentation. The student leaving school with expertise in these systems is prepared for the multidisciplinary approaches to therapeutics and diagnostics. Mass spectrometry is being used today for a broad range of applications including determining the change in protein expression in humans between disease and normal states. In addition, the quantitation of proteins by this technique and additional analytical technology is also critical to future medicines.

Those undergraduate students who seek to become analytical scientists will benefit from attending a graduate school with a strong and diverse program in analytical chemistry. Young scientists have an opportunity for active learning by working in a strong research program and from supportive and informative interactions with peers. The top three schools in *US News and World Report* annual rankings (*6*) of graduate chemistry programs for 2006 are University of North Carolina at Chapel Hill, Purdue University and University of Illinois, Champaign-Urbana. All three schools are recognized not just for their staff's research but also their links to industry; professors often collaborate or join forces with companies in the development of state-of-the-art analytical instruments or technology.

Internships and Co-ops

Some knowledge required for industrial careers may not be gained in the typical university curriculum; it can be obtained, for example, through an industrial internship or co-operative education program (Co-op). Internships in industry during the undergraduate and graduate period are a vital form of active learning. Whether an undergraduate student plans to go directly to industry or to pursue an advanced degree there is benefit to spending at least one summer in an industrial position. The student learns to use the latest equipment and to work in teams. The student also begins to understand the industrial world where flexibility in approaches to problems and to ones career path is more important than in the academic community. Co-op periods form the cornerstone of some university programs and industry-university relationships.

Boeing has a Co-op program where students taking the senior design course tackle a real world problem and split their time between industry and the classroom (*7*). The class concentrates on broad concepts though the projects may be part of a real problem in the company. Students also gain insight into real industrial applications and an understanding that, while top students excel individually, it takes the whole team to make a touchdown. Boeing has hired 25 percent of the current class of twenty-two students. The advantage of hiring these students is that Boeing already knows how they work and how they react under pressure. Learning the process and being adaptable are keys for a good player whether on the football field or on the job. The material actually learned is just a first down toward the goal.

Conferences and Conference Courses

Future analytical chemists can also enhance their preparation for industry by taking courses at conferences. An example is the course "Professional Analytical Chemists in Industry." Proctor & Gamble Company (P&G) sponsors the course and the instructors are drawn from P&G scientists (*8*). This course was given at such conferences as FACCSS (www.facss.org) and Eastern Analytical Symposium (www.eas.org) in 2005 and at Pittcon 2006 (www.pittcon.org). The program was intended primarily for undergraduate students, to educate them about careers as analytical chemists in industry, and to teach problem-solving skills. The focus was extensively on utilizing real problems from P&G. Students come away with additional skills, including a "framework" for approaching problems. They were also made aware of the different roles of the industrial analytical chemist which included scientific consultant, methods developer, and problem solver. In addition to the obligatory good communications skills and the ability to solve problems, students taking these types of courses will be better equipped to weather the challenges to livelihood in 2015.

Continuing Education and Skill Strengthening in Industry

Continuous on-the-job training is critical regardless of the educational background that the scientist has received prior to entering the workforce. What are the formats for this training?

The industry places a heavy emphasis on continuing education for employees, and many firms provide classroom training in safety, environmental and quality control. Some pharmaceutical companies offer training programs to help scientists and engineers keep abreast of new developments in their fields and to develop administrative skills (*5*). These programs may include meetings and seminars with consultants from various fields. Many companies encourage scientists and engineers to further their education; some companies provide financial assistance or full reimbursement of expenses for this purpose. Most newly employed technical sales representatives complete rigorous formal training programs revolving around their company's product lines.

State-of-the art instrument manufacturers offer training on their systems and the technology behind these products. For example, Waters Corporation and Applied Biosystems websites describe their customer training programs (www.waters.com and www.appliedbiosystems.com). The educational programs for both companies provide knowledge and skills for both those new to separation technology and systems as well as experienced professionals. Both companies offer comprehensive, hands-on instrument training sessions as well as applications-based courses on a variety of topics of interest to analytical chemists. Each company teaches courses that range from their new technologies and instrumentation to applications and information management solutions. They provide scientists with LC and MS knowledge and the expertise often required to be successful. They offer courses at their training centers around the world. Waters also staffs on-site education that fits a company's specific needs.

Conference for the Industrial Scientists

While conference courses were mentioned as part of the educational experience prior to joining industry, afterward they become a major training tool available to analytical chemists. These courses normally follow the trends in the chemical enterprise thus keeping the scientist up to date in new technology and/or applications. Among the courses offered recently by ACS (www.chemistry.org) for analytical chemists are

- Fourier Transform Infrared Spectroscopy: A Hands-on Workshop;
- IR Spectral Interpretation: A Systematic Approach;
- Interpretation of Mass Spectra;
- LC/MS: Fundamentals and Applications;
- Peptide and Protein Characterization with Mass Spectrometry.

Not only the courses but the conferences themselves serve a vital function in active learning for the analytical chemists. Many skills are strengthened by participating in technical meetings. Scientists discuss their work by oral lectures and poster presentations as well as during informal dialog. Publications can also result from giving a lecture or poster at a conference; several technical meetings issue specific journal volumes containing the papers that were presented. Thus, speaking and writing skills grow which are keys to success in industry. The global chemical enterprise is also reflected in the location of these meetings. An industrial chemist should attend and present not only in the United States and Western Europe but also in China and India. Both locations hold conferences in English and also give the attendees a glimpse into the new job market.

The Human Proteomics Organization (HUPO) annual congress speaks to both the new multidisciplinary programs and the global enterprise (www.HUPO.org) . HUPO's third world conference in 2004 was held in Beijing, China with worldwide attendance. The congress drew an exhibition from HUPO's industrial partners in proteomics as well as vendors in clinical diagnosis, therapeutics, and pharmaceutics on their new products, novel technologies, and new prognostic and diagnostic tools. The plenary lectures addressed such interdisciplinary topics as proteomic databases, deciphering human proteomes with mass spectrometry, nanotechnology, antibody arrays, systems biology and biomarker validation. The education of analytical chemists in industry should include such global exposure and diverse programs.

Preparing for Non-Laboratory or Hands-Off Roles

Industry often gives analytical chemists an option to pursue either the scientific or managerial ladder as they move forward in their careers. Those chemists choosing or chosen for the managerial ladder often attend specific management training courses. Several companies and business school offer these courses; analytical chemists with and without graduate degrees may decide to get an MBA to improve their non-technical skills. For example, when scientists transition to marketing, they should consider enrolling in a basic class on this discipline. Analytical chemists moving to project managers can follow a specific curriculum offered by schools such as Worcester Polytechnic Institute (WPI). The need for additional training continues as one climbs the corporate ladder but the basic skills obtained early in ones career remain essential.

Summary

The analytical chemists of 2015 and beyond have many opportunities if they are prepared. The universities do provide a good foundation in basic science, but they must also make sure that the future chemists have the opportunity to develop skills and expertise in state of the art technology. In addition, the student has to be active in supplementing his academic program with internships and conference programs. The development of writing and oral communication as well as problem-solving skills should start at the undergraduate level and continue throughout ones career. While pursuing an industrial career, learning continues as one bridges the gap created by moving up the ladder as a scientist or a manager.

References

1. Frantz, S. *Nature Rev. Drug Discovery* **2003**, *2*, 243. http://www.nature.com/drugdisc/nj/articles/nrd1058.html (accessed December 7, 2006)
2. American Chemical Society. American Chemical Society 2005 Annual Report. http://www.chemistry.org/portal/a/c/s/1/acsdisplay.html?DOC=acsinfo%5Cannualreport.html (accessed December 7, 2006)
3. William F. Carroll, Jr. and Douglas J. Raber. The Chemical Enterprise 2015. http://www.chemistry.org/portal/resources/ACS/ACSContent/industry/chementerprise2015/chem_enterprise.pdf (accessed December 7, 2006)
4. Petkewich, R. *C&E News* **2005**, *83*, 45.
5. DiversityWorking.com, Pharmaceutical and Medicine Manufacturing. http://www.diversityworking.com/career/Pharmaceuticals/Pharmaceutical_and_Medicine_Manufacturing/Analytical_chemist.html (accessed December 7, 2006)
6. U. S. News and World Report. U.S. News and World Report Ranking of Graduate Schools for 2007. http://www.usnews.com/usnews/edu/grad/rankings/rankindex_brief.php (accessed December 7, 2006)
7. Students Tackle Boeing Design Problem. http://www.eng.iastate.edu/muses/fall97/boeing.htm (accessed December 7, 2006)
8. Proctor & Gamble. Diverse Careers in Technology, Professional Analytical Chemists in Industry Course. http://www.pg.com/science/prof_chemists.jhtml;jsessionid=3X5TB0NKMIBOVQFIAJ1S0HWAVABHMLKG (accessed December 7, 2006)

Chapter 3

Education for Analytical Chemistry in the United States from the 1950s to the Present

Frank A. Settle, Jr.

Department of Chemistry, Washington and Lee University, Lexington, VA 24450

This chapter provides an overview of some important developments in the analytical curriculum in the United States from the 1950s to the present. It emphasizes problem based learning as an important component of courses and laboratories in this curriculum. It also points out how courses, laboratories, and textbooks have changed to encompass advances in instrumentation, computers, content, and pedagogy. Finally the author suggests unique contributions that analytical chemistry can offer in the training of current and future scientists and engineers.

Introduction

The ability to determine the composition of diverse materials is critical to all areas of chemistry and related fields of science. This qualitative and quantitative characterization, which established analytical as a sub-discipline of chemistry, is based on practical, hands-on methodology. Thus, effective instruction in analytical chemistry has included problem-based learning (PBL) and applied laboratory experiences.

In the early part of the last century the undergraduate chemistry curriculum in United States, reflecting a strong European influence, consisted of four sub disciplines; inorganic, analytical, organic, and physical presented in that order. Analytical was taught as an applied course in inorganic qualitative analysis that was followed by a rigorous study of quantitative analysis that emphasized gravimetric procedures. Students in advanced courses then applied these procedures and techniques to diverse inorganic samples including ores, rocks, metals, and alloys.

The introduction of volumetric analysis with its associated titrimetric methods into the quantitative course occurred soon after. Later on potentiometric and conductometic methods augmented visual endpoint indicators. Emphasis on fundamental principles included inorganic solution equilibrium, precipitation, reaction kinetics, and preparation of standard solutions. From the 1930s through the 1950s successive editions of the text by Kolthoff and Sandell (*1*) that emphasized these fundamentals greatly influenced the content of these courses. Inorganic analysis continued to be stressed to the exclusion of organic analysis which was confined to organic courses. Many schools included a course in organic qualitative analysis in addition to the introductory organic course. This division of material in the curriculum remained the status quo until the 1950's.

During this period spectrochemical and electrochemical methods began to appear in advanced analytical courses (usually titled instrumental analysis) and over time these courses became well established often taking the place of the second semester of quantitative analysis. The American Chemical Society's Committee on Professional Development recognized this trend in the 1950s and decreased the requirement for the introductory quantitative analysis course from a year to a semester followed by a one semester advanced analytical course for which physical chemistry was a prerequisite.

Prof. Herbert Laitinen (*2*) observed that either classical analytical chemists or electrochemists usually developed the advanced analytical courses. Spectrometry remained in the domain of the physical chemistry. In the 1950s, the advent of gas chromatography underscored the importance of separation science in chemical analysis. The convergence of electrochemical, spectrometric, and chromatographic methods, along with a more systematic study of instrumentation, influenced the content of instrumental analysis courses and led to the emergence of analytical chemistry as a robust sub-discipline of the modern chemistry curriculum.

Advances in commercially available instrumentation required to support undergraduate laboratory courses often included instruments that were affordable and well suited for the undergraduate laboratory. Examples included the Perkin-Elmer 137 infrared spectrometer, the Beckman DU UV-visible spectrometer, the Bausch and Lomb Spectronic 20 visible spectrometer, a variety of pH meters and electrodes, the Sargent polarographs, and the Gow Mac gas chromatograph. Many of these instrumental techniques were also used to support undergraduate research projects. Funding for undergraduate instrumentation became available through private foundations such as the Research Corporation and public agencies, primarily the National Science Foundation. One particularly successful program was the NSF's Instrumentation and Laboratory Improvement (ILI) Program which provided instrumentation for undergraduate laboratories in hundreds of colleges and universities from 1985 to the late 1990s. A number of instrumental textbooks also became available to support the introduction of instrumentation into the chemistry curriculum.

However, as Prof. Laitinen (*3*) observed in the 1970s, the analytical instructor was caught between presenting the classical methods of quantitative analysis, fundamental to chemical analysis, and introducing modern instrumental techniques with applications in many areas of chemistry and allied disciplines. If she placed too much emphasis on the classical methods, the modern techniques would often appear in courses outside the analytical curriculum and her course would become dated. On the other hand, the introduction of instrumental methods reduced the time devoted to basic principles. An answer to this quandary was the introduction of simple instruments such as visible spectrometers and pH meters into the introductory quantitative analysis course. Thus, spectrophotometric and potentiometric titrations became common components of the laboratory, replacing the multi-component gravimetric analysis of brass or limestone that were staples of the earlier classical laboratory. Professor Royce Murray (*4*) observed that a shift in emphasis on titrimetry from means of analysis to a method of illustrating the fundamentals of spectrophotometric and electrochemical detectors, coordination chemistry, and chemical equilibrium, ensured the place of titrimetry in both advanced and introductory courses.

Curriculuar Developments

In 1970, a series of articles on education appeared in the A-pages of *Analytical Chemistry.* These articles reflected the current state and future of analytical chemistry in the chemistry curriculum. Prof. Sidney Sigga (*5*) at the University of Massachusetts, Amherst, noted that a good instructor designed a course that struck a balance between each of the following; modern and classical material, theory and application, method development and detailed procedures. He also suggested that the enthusiasm of the instructor was a critical component in successful courses. Prof. Laitinen (*6*) emphasized non-traditional approaches to teaching chemical analysis. His suggestions included the introduction of improved qualitative methods (spot tests, flame analysis, paper chromatography, etc.) throughout the chemistry curriculum to introduce student to analysis, use of instruments into introductory quantitative analysis courses, insertion of analytical type experiments in physical chemistry laboratories, and sharing of research instruments with instructional laboratory courses. A third article by Roland Hirsch (*7*) suggested fundamentals to be presented in analytical courses. These included importance of the analyte signal, experiment and instrument design, quantifiable principles (equilibra, kinetics, etc.), stoichiometry, and mini-research problems.

Prior to 1960, students wishing to acquire a working knowledge of instrumentation had to take courses in advanced physics or electrical engineering which had numerous prerequisites. As a result few chemistry students acquired an in-depth knowledge of instrumentation. In the early 1960s, Prof. Howard Malmstadt developed a one-semester course at the University of Illinois that successfully introduced students to the electronic components of instrumentation by enabling them to build their own instruments. His book (*8*), *Electronics for Scientists*, co-authored with Profs. Chris Enke and E. Clifford Toren, Jr., and a series of NSF sponsored short courses introduced a generation of instructors and students to the electronics of instrumentation. Enke also developed the concept of data domains (*9*), a useful paradigm that assisted students in following analyte information from chemical and physical environments through electronic transformations to useful output. Later as digital electronics and computers entered the laboratory, either as stand alone devices or as instrument components, Prof. Sam Perone, at Purdue, developed a popular course and accompanying text (*10*) in interfacing computers to instruments. This course expanded on Prof. Malmstadt's approach to the use of computer input/output signals for data acquisition and instrument control. Both courses were supported by components manufactured by the DC Heath Company which provided an active learning environment where students assembled instruments from components and interfaced them to laboratory computers to perform chemical analyses. This generated much excitement for students and professors alike.

Courses continued to evolve in the next twenty years with the advent of new instrumentation and methodology. Qualitative analysis moved from a separate sophomore-level course to the general chemistry laboratory. Advanced courses removed the covers from instruments to focus on components and their functions. Calculators and computers aided in the statistical analysis of data. This period also saw an increased emphasis on bioanalytical applications, many of which involved elements of separation science. A sophomore quantitative course that represented many of these trends included access to contemporary instrumentation (HPLC, voltametry, atomic absorption, and potentiometry with ion selective electrodes), incorporation of *Good Laboratory Practices*, (*11*) realistic sampling, inclusion of analytical problems from many areas (biological, environmental, etc.) and multi-element analysis of mixtures (*12*).

Use of PBL in Class and Laboratory

Traditionally, the laboratory involved the analysis of "unknown" materials dispensed by the instructor from bottles of Thorn Smith's pre-analyzed samples. While this procedure developed good fundamental procedures and techniques, it is rather sterile in that it often denied the student a view of the overall problem to be addressed by the analysis. Students who took samples from a commercial product or collected them from an environmental site were more likely to "own" them as opposed to students receiving a pre-analyzed sample dispensed by the instructor. This "ownership" caused students to become more engaged in the analysis and developed a comprehensive view of the problem. In 1971, Prof. Earl Wehry (*13*) described a laboratory that included 25 mini-research experiments whose scope was broad enough to give student options for analyses, used simple instrumentation, required error detection and control and represented real world problems.

Another development was the application of problem-based learning (PBL) which appeared in general education pedagogy in the 1980s. Prof. John Walters (*14*) was one of the first to implement PLB in his analytical laboratory course at St. Olaf's College in the early 1990s. In this lab, groups of four students formed companies, where each student assumed the role of manager, chemist, hardware engineer, or software programmer. The companies then addressed laboratory problems known as "management dilemmas" with the students performing the functions of their assigned positions. This approach, while successful, required a well-equipped laboratory, an enthusiastic, dedicated instructor, and relatively few students. Profs. George Wilson, Marc Anderson, and Craig Lunte (*15*) at the University of Kansas modified Walters approach for use with a larger number of students with more limited access to instrumentation and computers. The laboratory course emphasized problem solving while the lecture component provided a survey of instrumentation appropriate for the laboratory. Students were "employed" by "companies" to work in teams to address real world problems and were guided by "managers" (graduate assistants), upper level management (course instructors) and "consultants" (industrial or government scientists who work with the course). The teams prepared both written and oral interim and final reports as well as poster presentations. In both cases, the key to the success PBL was the selection of interesting and challenging problems for the students.

Prof. Tom Wenzel (*16*) at Bates College made extensive use of PBL in both the classroom and laboratory. In this environment, lectures were replaced with group learning which involved active student participation in the preparation and presentation of materials. The course was restructured into analytical fundamentals and instrumental analysis. Project based laboratories involved teams of students who developed methods to solve semester long analytical problems. This approach was suited for smaller classes but involved large time commitments by both students and the instructor. Another interesting PBL exercise was "Project Select" implemented by Prof. Frank Settle (*17*) in which students were asked to select appropriate instrumental technique(s) to solve a analytical problem using information from available sources including manufacturers' literature. Students explained the fundamentals of the chosen technique(s) and the reasons for their choice(s) including specifications and limitations of the instrument(s).

Two other PBL laboratory courses focused on a single analyte. Prof. Alanah Fitch (*18*) at Loyola University used lead determinations as the basis for a course which involved students interacting with Chicago communities to obtain samples containing lead and prepare them for analysis by multiple techniques including UV-visible spectrometry, fluorescence quenching spectrometry, graphite furnace atomic spectrometry, FT-NMR spectrometry, cyclic voltametry, anodic stripping voltametry and ion selective potentiometry. The results were presented to the communities where the samples originated. Profs. Frank Settle and Michael Pleva (*19*) developed a semester long laboratory at Washington and Lee University that featured the determination of sodium in packaged foods. It made extensive use of experimental design and statistical analysis to compare variances in sampling, sample preparation, and measurement. In addition it also allowed the students to compare results from different techniques including atomic absorption spectrometry, atomic emission spectrometry, inductively coupled plasma emission spectroscopy, ion chromatography, and potentiometry (sodium ion electrode).

In addition to the application of computers as components of instrumentation, the availability of software influenced analytical education (*20*). Students could learn by developing BASIC or FORTRAN programs that involved equilibria, pH, and statistical problems. Later spreadsheets and more sophisticated program packages such as Math CAD and SPSS provided enhanced capabilities for both classroom and laboratory exercises. A number of simulators also appeared which allowed students to gain experience with operation and data interpretation of instruments that were not readily available. These included NMR, HPLC, IR, and electronic circuit design. A different simulation presented analytical chemistry in the context of an environmental problem facing a community with an industrial plant that produced BCTC, a hypothetical chemical (*21*). In this exercise students obtained samples, performed analyses, examined epidemiological data, and looked at economic impacts to access any problems due to pollution and if necessary, provide appropriate solutions.

Recent Developments

Prof. Murray (*22*) has observed that the course of education in analytical chemistry can be traced through the evolution of textbooks. As mentioned above, from 1936 to the 1960s the popular editions of Kolthoff and Sandell's quantitative analysis text (*23*) addressed the fundamentals of chemical reactivity and physical phenomena underlying gravimetric metric and titrimetric analysis in contrast to previous texts which focused on empirical discussions of techniques and procedures. The increased availability of instrumentation and its influence on analytical chemistry was reflected in the textbooks of the 1950s. By this time the contribution of analytical to the chemistry curriculum consisted of an "introductory" and an "advanced" course. Some, like Laitinen's *Chemical Analysis, An Advanced Text* (*24*), continued the focus on fundamentals following Kolthoff and Sandell. Others authors such as those by Delahay (*25*) and Ewing (*26*) emphasized instrumentation to the exclusion of gravimetric and titrimetric topics.

A 2000 survey of instrumental analysis courses (*27*) concluded that these courses continued to evolve with the content reflecting the concerns and sophistication of society. Topics once presented at the graduate level appeared in upper level undergraduate courses and topics from these advanced undergraduate courses moved to the sophomore analytical courses. The expansion of material in introductory texts such as Harris (*28*) reflected this trend. Instructors continued to be faced with choosing which topics are fundamental and necessary and those which are optional. The survey found that more than twice as many topics are covered in the lecture portion than in the laboratory. This is likely a function of availability of time and instrumentation.

In addition to textbooks, features in the A pages of *Analytical Chemistry* as well as articles and features in the *Journal of Chemical Education* (*JCE*) provided a record of trends in analytical education. Prof. Wenzel has edited a series of a page articles devoted to education in the a-pages since 1998. Profs. Ewing and Settle edited a feature on instrumentation in *JCE* from 1966 to 1987. In addition *JCE* contains numerous articles on the content, presentation, and evaluation of analytical courses.

In 1996-97 the Division of Undergraduate Education and the Chemistry Division of NSF sponsored a series of workshops that focused on curricular developments in the Analytical Sciences organized by Prof. Ted Kuwana of the University of Kansas. Participants included representatives from two- and four-year colleges, graduate-level institutions, and industry as well as government agencies and laboratories. After agreeing on the importance of measurement, problem solving, and hands-on experiences techniques in undergraduate courses, they focused on ways to bring real-world applications and problem-solving into the curriculum. The summary report (*29*) highlighted six priorities for the undergraduate curriculum: 1) course content and modular presentation, 2) core technologies for laboratory work, 3) faculty development, 4) learning partnerships with industry, 5) integration of modern technology, and 6) follow-up activities and dissemination.

One of the results of these meetings was the development and implementation of the Analytical Sciences Digital Library (*30*), a peer-reviewed, web-based resource for persons with an interest in the applications of analytical chemistry. The target audience includes educators, students and practicing professionals who are involved with applications of analytical chemistry. The Library on-line since 2002 is a component of the NSF's National Science Digital Library. It contains information on instrumentation, applications, educational materials, and laboratory exercises organized to facilitate access to these materials.

A 2002 survey of undergraduate quantitative analysis courses by Prof. Patricia Mabrouk (*31*) revealed that substantive changes were occurring in these courses with respect to both content and presentation. Faculty expressed concern that students understand the science and technology behind modern instrumentation lest they become mere button pushers. There was also a concern that a balance be maintained between fundamentals, such as chemical equilibrium, and instrumentation. Nearly half of the respondents employed problem based learning in their courses.

The conclusions of a recent survey conducted by Prof. Julian Tyson and Ms. Angela Fahey (*32*) indicated a number of interesting developments. These include the appearance of nuclear magnetic resonance spectroscopy and mass spectrometry, formerly presented in organic courses, in the analytical curriculum and the convergence of instrumental analysis and traditional quantitative courses. The authors suggest that these trends may be influenced by the integration of biological topics into the curriculum and by the overage of instrumentation by other areas of chemistry. The survey also indicated that approximately 75% of the students encountered instrumentation in courses other than analytical chemistry.

The Future Role of Analytical Education

Education for analytical chemistry will continue to evolve with changes in the discipline. Methodology and techniques associated with the life sciences plays a dominate role in many analytical courses. The analysis of interfaces between physical phases is another important area to be addressed. The analytical chemistry has always faced the paradox of a sub-discipline that supports many areas of science and technology. Practitioners in these areas have the ability to perform their own analysis, especially with the advent of modern, computerized instrumentation, so what can analytical chemistry contribute? The lids on the black boxes of current instrumentation have been tightened by the extensive use of built-in computers for control as well as data acquisition and processing. In this environment analytical chemistry offers a methodology that includes a unique combination of experimental design, sampling, sample preparation, knowledge of the capabilities and limitations of measurement techniques, data analysis, and presentation of results including validation procedures. These combine to provide students with real PBL experiences, the most important components of education for chemical analysis.

The Author's Personal Reflections

Since the 1950s many people have contributed to the development of the analytical curriculum. In addition to Prof. Malmstadt, the following five persons greatly influenced the content of the author's undergraduate courses and laboratories. In additional to their interest in education, each was generalist with research interests in diverse areas. Dr. John Taylor, a chemist at the National Institute of Standards and Technology influenced generations of professors, students, and practicing professionals through his publications and short courses addressing sampling, experimental design, data analysis, and quality assurance. Prof. Sidney Sigga at the University of Massachusetts, whose early career as an industrial chemist led him to develop a practical approach to chemical analysis, was able to incorporate the pragmatic methods of an industrial environment into the analytical curriculum. His *Survey of Analytical Chemistry* (*33*), published in 1968, combined common sense with the use of chemical reactions and a full range of instrumental techniques to develop methods of analysis. Prof. Lockhart "Buck" Rogers, an eminent researcher whose career included positions in leading universities, that included MIT, Purdue, and the University of Georgia, also promoted the education of numerous graduate and undergraduate students through the combination of instruction and research. Prof. John Dean at the University of Tennessee led the publication of the initial textbook on instrumental analysis in 1945 (*34*) and continued to oversee it through seven editions. This text supported the initial courses in instrumental analysis courses and spawned a number of other texts in the field. Finally, Prof. Charles Reilley at the University of North Carolina and Prof. Donald Sawyer at the University of California Riverside, authored *Experiments for Instrumental Methods* (*35*), one of the first laboratory manuals that had a major impact on the instrumental laboratory.

References

1. Kolthoff, I. M.; Sandell, E. B. *Textbook of Quantitative Inorganic Analysis, 1st edition*; The MacMillian Co., New York, NY, 1936.
2. Laitinen, H. A. *Z. Anal. Chem* **1973**, *72,* 309.
3. Laitinen, H. A. *Z. Anal. Chem* **1973**, *72,* 310.
4. Murray, R. W. *Talanta* **1989**, *36,* 13.
5. Sigga, S. *Anal. Chem* **1970**, *42,* 49A-52A
6. Laitinen, H. A. *Anal. Chem* **1970**, *42,* 37A-41A.
7. Hirsch, R. F. *Anal. Chem* **1970**, *42,* 42A-48A.
8. Malmstadt, H. V., Enke, C. G., and Toren, E. C.; *Electronics for Scientists: Principles and Experiments for those Who Use Instruments*; W. A. Benjamin: New York, NY, 1962.
9. Enke, C. G., *Anal. Chem.* **1971**, *43,* 69A.
10. Perone, S. P. and Jones, D. O *Digital Computers in Scientific Instrumentation: Applications to Chemistry* ; McGraw-Hill, New York, NY, 1973.
11. U.S. Food and Drug Administration, Office of Regulatory Affairs. http://www.fda.gov/ora/compliance_ref/bimo/glp/default.htm (accessed November 2006).
12. Perone, S. P., Pesek, J., Stone, C, Englert, P.; J. Chem. Ed. **1998**, *75*, 1444-1452.
13. Wehry, E. L.; J. Chem. Ed. **1970**, *47*, 843-844.
14. Walters, John P. *Anal. Chem.* **1991**, *63,* 977A-985A; *Anal. Chem.* **1991**, *63,* 1077A-1084A; *Anal. Chem.* **1991**, *63,* 1179A-1191A.
15. Wilson, George S.; Anderson, Mark R.; Lunte, Craig E. *Anal. Chem.* **1999**, *71*, 677A-681A.
16. Wenzel, T. *Anal. Chem* **1998**, *70,* 790A-795A.
17. Settle, F. *J. Chem. Educ.* **1979**, *56*, 411-412.
18. Fitch, A.; Wang, Y.; Mellican, S; Macha, S *Anal. Chem.* **1996**, *68,* 727A-731A.
19. Settle, F. A.; Pleva, M. A *Anal. Chem.* **1999**, *71*, 538A-540A, 727A-731A.
20. O'Haver, T. C. *Anal. Chem.* **1991**, *63*, 521A-526A.
21. Whisnant, D. M.; McCormick, J. A. *J. Chem. Educ.* **1999**, *76*, 719.
22. Murray, R. W. *Talanta* **1989**, *36,* 11-17.
23. Kolthoff, I. M.; Sandell, E. B. *Textbook of Quantitative Inorganic Analysis, 1st edition*; The MacMillian Co., New York, NY, 1936.
24. Laitien, H. A. *Chemical Analysis: An Advanced Text and Reference,* McGraw-Hill, New York, NY, 1960.
25. Delahay, P. *Instrumental Analysis;* The MacMillian Co., New York, NY, 1957

26. Ewing, G. *Instrumental Methods of Analysis*; McGraw-Hill, New York, NY, 1954.
27. Girard, J. E.; Diamant, C. *J. Chem. Ed.* **2000**, *77*, 646-648.
28. Harris, D. C. *Quantitative Chemical Analysis, 5th edition*; W. H. Freeman and Co., New York, NY, 1999.
29. Kuwana, T.; Curricular Developments in the Analytical Sciences. http://www.chem.ku.edu/TKuwana/CurricularDevelopment/ (accessed 10/19/2006).
30. Larive, C. K.; Kuwana, T.; Chalk, S. *Anal. Chem.* **2004**, *76*, 398A-402A.
31. Mabrouk, P. *Ana. Chem.* **2002**, *74*, 268A-274A.
32. Tyson, J.; Fahey, A. *Anal. Chem.* **2006**, *78,* 4249-4254.
33. Siggia, S. *Survey of Analytical Chemistry* ; McGraw-Hill, New York, NY, 1968.
34. Willard, H. H.; Merritt, L. L., Jr.; Dean, J. A. *Instrumental Methods of Analysis, 1st edition* ; D Van Nostrand Co., New York, NY, 1945.
35. Reilley, C. N.; Sawyer, D. T *Experiments for Instrumental Methods: A Laboratory Manual;* McGraw-Hill, New York, NY, 1961

Chapter 4

Cooperative Learning

Richard M. Felder[1] and Rebecca Brent[2]

[1]Department of Chemical Engineering, North Carolina State University, Raleigh, NC 27695–7905
[2]Education Designs, Inc., 101 Lochside Drive, Cary, NC 27518

The term *cooperative learning* (CL) refers to students working in teams on an assignment or project under conditions in which certain criteria are satisfied, including that the team members be held individually accountable for the complete content of the assignment or project. This chapter summarizes the defining criteria of cooperative learning, surveys CL applications, summarizes the research base that attests to the effectiveness of the method, and outlines proven methods for implementing CL and overcoming common obstacles to its success.

Introduction

Many students who have worked in a team in a laboratory- or project-based course do not have fond memories of the experience. Some recall one or two team members doing all the work and the others simply going along for the ride but getting the same grade. Others remember dominant students, whose intense desire for a good grade led them to stifle their teammates' efforts to contribute. Still others recall arrangements in which the work was divided up and the completed parts were stapled together and turned in, with each team member knowing little or nothing about what any of the others did. Whatever else these students learned from their team experiences, they learned to avoid team projects whenever possible.

Cooperative learning is an approach to groupwork that minimizes the occurrence of those unpleasant situations and maximizes the learning and satisfaction that result from working on a high-performance team. A large and rapidly growing body of research confirms the effectiveness of cooperative learning in higher education (*1-4*). Relative to students taught traditionally—i.e., with instructor-centered lectures, individual assignments, and competitive grading—cooperatively taught students tend to exhibit higher academic achievement, greater persistence through graduation, better high-level reasoning and critical thinking skills, deeper understanding of learned material, greater time on task and less disruptive behavior in class, lower levels of anxiety and stress, greater intrinsic motivation to learn and achieve, greater ability to view situations from others' perspectives, more positive and supportive relationships with peers, more positive attitudes toward subject areas, and higher self-esteem. Another nontrivial benefit for instructors is that when assignments are done cooperatively, the number of papers to grade decreases by a factor of three or four.

There are several reasons why cooperative learning works as well as it does. The idea that students learn more by doing something active than by simply watching and listening has long been known to both cognitive psychologists and effective teachers (5, 6) and cooperative learning is by its nature an active method. Beyond that, cooperation enhances learning in several ways. Weak students working individually are likely to give up when they get stuck; working cooperatively, they keep going. Strong students faced with the task of explaining and clarifying material to weaker students often find gaps in their own understanding and fill them in. Students working alone may tend to delay completing assignments or skip them altogether, but when they know that others are counting on them, they are motivated to do the work in a timely manner.

The proven benefits of cooperative learning notwithstanding, instructors who attempt it frequently encounter resistance and sometimes open hostility from the students. Bright students complain about begin held back by their slower teammates; weak or unassertive students complain about being discounted or ignored in group sessions; and resentments build when some team members fail to pull their weight. Knowledgeable and patient instructors find ways to deal with these problems, but others become discouraged and revert to the traditional teacher-centered instructional paradigm, which is a loss both for them and for their students.

In this chapter we describe cooperative learning methods that have been proven effective in a variety of instructional settings. We then suggest ways to maximize the benefits of the approach and to deal with the difficulties that may arise when cooperative learning is implemented.

What is Cooperative Learning?

Several definitions of cooperative learning have been formulated. The one most widely used in higher education is probably that of David and Roger Johnson of the University of Minnesota. According to the Johnson & Johnson model, cooperative learning is instruction that involves students working in teams to accomplish a common goal, under conditions that include the following elements (*7*):

1. **Positive interdependence.** Team members are obliged to rely on one another to achieve the goal. If any team members fail to do their part, everyone suffers consequences.

2. **Individual accountability.** All students in a group are held accountable for doing their share of the work and for mastery of all of the material to be learned.

3. **Face-to-face promotive interaction.** Although some of the group work may be parcelled out and done individually, some must be done interactively, with group members providing one another with feedback, challenging reasoning and conclusions, and perhaps most importantly, teaching and encouraging one another.

4. **Appropriate use of collaborative skills.** Students are encouraged and helped to develop and practice trust-building, leadership, decision-making, communication, and conflict management skills.

5. **Group processing.** Team members set group goals, periodically assess what they are doing well as a team, and identify changes they will make to function more effectively in the future.

Cooperative learning is not simply a synonym for students working in groups. A learning exercise only qualifies as cooperative learning to the extent that the five listed elements are present.

Cooperative Learning Structures

Cooperative learning can be used in for any type of assignment that can be given to students in lecture classes, laboratories, or project-based courses. Following are some of the structures that have been used, with some recommendations for how they may be effectively implemented. (Additional suggestions are given at the conclusion of the chapter.)

Problem Sets

Students complete some or most of their homework assignments in teams. The teams are encouraged to include only the names of actual participants on the solution set that they hand in. The students are initially disinclined to leave anyone's name off, but eventually they get tired of letting nonparticipants ("hitchhikers," in cooperative learning parlance) get good grades for work they didn't do and begin to omit names, at which point many hitchhikers—unhappy about getting zeroes on assignments—start cooperating.

The team gets a grade for the assignment, but eventually the performance of each team member should be assessed and the results used to adjust the average team homework grade separately for each team member. Adjusting team grades for individual performance is one of the principal ways of assuring individual accountability in cooperative learning, second only in importance to giving individual exams. Later in this chapter we will describe systems for performing the performance assessments and making the adjustments.

We recommend using a mixture of individual and team assignments in a lecture course rather than having all assignments completed by teams. One obvious reason is to provide another measure of individual accountability. Another is that if there is a lot of dropping and adding in the first one or two weeks of the course, it is better to wait until the class population stabilizes before forming teams.

We also suggest advising teams not to simply meet and complete each assignment together. One team member is usually the fastest problem solver and begins almost every homework problem solution in the group sessions, and the other members then have to figure out how to get the solutions started for the first time on the individual tests, which is not a good time for them to have to do it. We recommend instead that all team members outline solutions individually before meeting to work out the details. On the first few assignments we require team members to sign and hand in their outlines to help them acquire the habit.

Laboratories and Projects

Laboratories and projects may be carried out by teams (as they often are in traditional curricula), except that again the team grades should be adjusted for individual performance.

The problem with team labs and projects as they are normally conducted is that there is no individual accountability at all. The result is the familiar situation in which some team members do the bulk of the work, others contribute little and understand little or nothing about the project, everyone gets the same grade, and resentment abounds. Adjusting the team project grades for individual performance goes a long way toward correcting these injustices. In addition, it is good practice to include some individual testing on every aspect of the project and have the results count toward the final course grade. If this is done, hitchhikers who understand either nothing or only the little they did personally will be penalized and perhaps induced to play a more active role in subsequent work.

Jigsaw

Jigsaw is a cooperative learning structure applicable to team assignments that call for expertise in several distinct areas. For example, in a laboratory exercise, areas of expertise might include experimental design, equipment calibration and operation, data analysis (including statistical error analysis), and interpretation of results in light of theory, and in a design project the areas might be conceptual design, process instrumentation and control, safety and environmental impact evaluation, and cost and profitability analysis.

Suppose four such areas are identified for a project. The students are formed into teams of four, and either the instructor or the team members designate which member will be responsible for each area. Then all the experts in each area are given specialized training, which may involve getting handouts or presentations by the course instructor, a faculty colleague, or a graduate student knowledgeable in the area in question. The students then return to their home teams and complete the assignment. The teams count on each member to provide his or her expertise, and if an expert does a poor job, the quality of the final project is compromised and everyone's grade suffers. Moreover, if the students are tested on all of the areas of expertise, the overall learning from the assignment improves dramatically. The tests require all students to understand the entire project, and not just the part that they were the experts in (individual accountability), and the experts have the responsibility of transmitting their expertise to their teammates (positive interdependence).

Peer Editing

When teams turn in written lab reports and/or give oral presentations, the usual procedure is for the instructor to do the critiquing and grading. A powerful alternative is *peer editing*, in which pairs of groups do the critiquing for each other's first drafts (written) or run-throughs (oral). The groups then revise their reports and presentations taking into account the critiquing teams' suggestions and then submit or present to the instructor. This activity lightens the grading load for instructors, who end up with much better products to grade than they would have without the first round of critiquing.

If a grading checklist or rubric is to be used for grading the team reports (which is always a good idea), it should be shared with the students before the reports are written and used for the peer editing. This practice helps the students understand what the instructor is looking for and invariably results in the preparation of better reports, and it also helps assure that the peer critiques are as consistent and useful as possible. If several rounds of peer editing are done and the instructor collects and grades the checklists or rubrics for the first one or two rounds, the students will end up giving much the same rubric scores as the instructor gives, and in good classes the instructor may only have to do spot checks of peer grades instead of having to provide detailed feedback on every report.

Peer-Led Team Learning

In *peer-led team learning* (PLTL), lectures are supplemented by weekly 2-hour *workshops* in which students work in six- to eight-person groups to solve structured problems under the guidance of trained peer leaders. The problems must be challenging and directly related to the course tests and other assessment measures. The course professor creates problems and instructional materials, assists with the training and supervision of peer leaders, and reviews progress of the workshops. The materials prompt students to consider ideas, confront misconceptions, and apply what they know to the solution process. The peer leaders clarify goals, facilitate engagement of the students with the materials and one another, and provide encouragement, but do not lecture or provide answers and solutions (*8, 9*).

PLTL was developed by chemistry educators in the 1990s and may be the most prominent group-learning strategy in chemistry education. (We will later describe illustrative implementations of the approach.) It is not a cooperative learning strategy by definition, but as Tien *et al.* (*10*) point out, it shares a number of elements with CL. The students are confronted with difficult problems and must rely primarily on one another to develop solutions, which promotes positive interdependence, and face-to-face interaction is crucial to the workshop format. Students are tested individually on the knowledge required to solve the problems, and a function of the peer leader is to get team members to explain their understanding to their teammates, both of which provide individual accountability. There is no formal instruction in teamwork skills in PLTL, but informal instruction invariably occurs as the peer leaders facilitate the group interactions. The only CL criterion that does not appear to be addressed as part of the PLTL model is regular self-assessment of team functioning, and it would be trivial to add that in PLTL implementations.

Applications in Chemistry Education

The literature of applications of cooperative learning in science, technology, engineering, and mathematics is quite large, and a comprehensive review of it is well beyond the scope of this chapter. We will confine ourselves here to describing several examples of applications in chemistry courses.

A bibliography assembled by Nurrenbern and Robinson (*11*) cites references to roughly 30 studies of team-based learning in chemistry lecture and laboratory courses, and a search for articles in recent issues of the *Journal of Chemical Education* that included cooperative learning among the key words revealed 47 articles published in 2004, 2005, and the first half of 2006. In the remainder of this section we describe several of these studies.

Hinde and Kovac (*12*) discuss two courses that introduced team-based learning in different ways. In the second semester of a physical chemistry course for chemistry and chemical engineering majors, biweekly computer-based group work sessions supplemented traditional lectures, and in the the second semester of a biophysical chemistry course taken primarily by biochemistry majors, an approach based on group work with occasional supplementary mini-lectures was used. The group sessions in both courses were inquiry-based. The self-selected teams of three or four in the biophysical chemistry course were given guidelines on effective teamwork, and both peer ratings and self-ratings of student performance on teams contributed to the final course grades. In the physical chemistry course there was little difference in performance between the class in question and previous classes that had been taught without group work, but this result is not surprising in view of the fact that the group activities were infrequent and most of the defining criteria for cooperative learning were not met. In the biophysical chemistry course the instructor's assessment was that the students gained considerable conceptual understanding and problem-solving ability as well as critical thinking and teamwork sklls, but no comparison with a control group was carried out that would elevate the assessment of the course beyond the anecdotal level. The author concludes that the course would have been improved by providing more structure and feedback, maintaining a better balance between individual and group work, and doing more to promote individual accountability (e.g., give more individual tests) and positive interdependence (e.g., establish and rotate assigned roles within teams).

A better example of cooperative learning implementation and assessment is provided by Tien *et al.* (*10*), who conducted peer-led team learning in a first-semester organic chemistry course over a three-year period and compared the performance of the students with the that of students who had taken a traditional version of the course in the preceding three years. The course instructor, text, examination structure, and grading system were the same for both the treatment and comparison groups. While instruction in teamwork skills is not necessarily a component of PLTL, in this case the peer leaders were trained in group dynamics and group skills and used their training to help the student teams learn to function effectively. It is therefore fair to say that the PLTL implementation described in this study fully qualifies as cooperative learning. On average, the workshop students significantly outscored their traditionally-taught counterparts on individual course exams, final course grades, retention in the course, and percentage earning the minimum acceptable grade of C– for moving on to the second semester organic chemistry course. Similar results were obtained specifically for female students and underrepresented minority students. The treatment group found the workshops and workshop problems their most important aids to learning in the course. Similar findings have been reported for PLTL programs in an organic chemistry class at another institution (*13*) and in a biology course (*14*), as well as for a cooperative learning implementation in organic chemistry (*15*).

A classical implementation of cooperative learning in chemistry is that of Hanson & Wolfskill (*16*), who used a "process workshop" format in the general chemistry class at SUNY-Stony Brook. Students worked in teams of three or four on activities that involved guided discovery, critical thinking questions that help provide the guidance, solving context-rich and sometimes open-ended and incompletely defined problems, and metacognitive reflecting. Most activities focused on a single concept or issue and could be completed in a 55-minute session. Following each workshop, students completed an individual quiz on the workshop content, thus promoting individual accountability. The use of this approach led to substantially improved examination grades relative to the previous year, in which the course was conventionally taught, as well as increased attendance at recitation and tutorial sessions and improvements in student self confidence, interest in chemistry, and attitudes toward instruction. The same authors report on an interactive computer-assisted learning model that supports and enhances the process workshop format by providing immediate feedback on student efforts, networked reporting capabilities, and software tools for both peer assessment and self-assessment (*17*).

Research Support for Cooperative Learning

Hundreds of research studies of team-based learning in higher education have been conducted, with most of them yielding positive results for a variety of cognitive and affective outcomes. Analyses of the research support the following conclusions:

- Individual student performance was superior when cooperative methods were used as compared with competitive or individualistic methods. The performance outcomes measured include knowledge acquisition, retention, accuracy, creativity in problem solving, and higher-level reasoning. Other studies show that cooperative learning is superior for promoting metacognitive thought, persistence in working toward a goal, transfer of learning from one setting to another, time on task, and intrinsic motivation For example, students who score in the 50th percentile when learning competitively would score in the 69th percentile when taught cooperatively (*1, 3*).

- Similar positive effects of group interactions have been found specifically for chemistry courses. In a meta-analysis of research on cooperative learning in high school and college chemistry courses, Bowen (*18*) found that students in the 50th percentile with traditional instruction would be in the 64th percentile in a cooperative learning environment.

- Several studies of active/collaborative instruction report positive effects on a variety of cognitive and affective outcomes. In a compilation of pre-post test gains in force concept inventory scores obtained by students in introductory physics courses, the use of instruction involving "interactive engagement" led to an average gain two standard deviations greater than was observed for traditionally-taught courses (*19*). Students in engineering capstone design courses taught with active and collaborative approaches outperformed traditionally-taught students in acquisition of design skills, communication skills, and teamwork skills (*4*). The use of collaborative methods had significant positive effects on understanding science and technology, analytical skills, and appreciation for diversity, among other outcomes (*20*).

- Affective outcomes were also improved by the use of cooperative learning. Relative to students involved in individual or competitive learning environments, cooperatively taught students exhibited better social skills and higher self-esteem (*3*), as well as more positive attitudes about their educational experience, the subject area, and the college (*7*). Towns *et al.* (*21*) used field notes and survey data to analyze students' attitudes toward group activities in a physical chemistry class. The students viewed the group work as a positive force in their learning, and they also valued the interactions for promoting a sense of community in the classroom.

Implementing Cooperative Learning

The benefits of using cooperative learning are well supported by theory and well established by classroom research, but the method is not without its problems, most of which have to do with individual student resistance and dysfunctional teams. Many techniques have been developed that minimize the problems, most of which involve addressing one or more of the five criteria for cooperative learning. The suggestions that follow are drawn primarily from Felder and Brent (*22, 23*), Johnson et al. (*7*), Oakley *et al.* (*24*) and Smith *et al.* (*2*).

Before we offer the suggestions, we should make clear that implementing cooperative learning successfully does not require adopting every one of them. In fact, trying to do so all at once might be a serious mistake: the instructor would have to juggle many unfamiliar techniques and end by doing none of them well, and the students would be deluged by an array of unfamiliar demands and many might rise up in rebellion. Rather, instructors new to cooperative learning should take a more gradual approach, choosing mainly the methods with which they feel most comfortable and adopting additional methods only when they have had time to get used to the current ones. If they do that, they will never stray too far from their comfort zones and will become increasingly adept at defusing student resistance long enough for the students to see the benefits of this new form of instruction for themselves.

Forming teams

Instructors should form teams rather than permitting students to choose their own teammates. When students self-select into teams, the best students tend to cluster, leaving the weak ones to shift for themselves, and friends cluster, leaving some students out of groups and excluding others from cliques within groups. Moreover, when graduates go to work in industry or business, they will be required to work in teams and will have no voice in the team formation, and their job performance evaluation will depend as much on their ability to work with their teammates as on their technical skills. Since that's what they'll be doing then, the job of their instructors is to prepare them for it now.

The following criteria are recommended for team formation:

1. *Form teams of* 3–4 *students for most tasks.* When students work in pairs, the diversity of ideas and approaches that leads to many of the benefits of cooperative learning may be lacking. In teams of five or

more, some students are likely to be inactive unless the tasks have distinct and well-defined roles for each team member.

2. *Make the teams heterogeneous in ability level.* The unfairness of forming a group with only weak students is obvious, but groups with only strong students are equally undesirable. The members of such teams are likely to divide up the homework and communicate only cursorily with one another, avoiding the interactions that lead to most of the proven benefits of cooperative learning. In heterogeneous groups, the weaker students gain from seeing how better students approach problems, and the stronger students gain a deeper understanding of the subject by teaching it to others.

3. *If the assignments require work being done outside class, form teams whose members have common blocks of time to meet during the week.*

4. *When students in a particular demographic category are historically at risk for dropping out, don't isolate members of that category in a team.* Students belonging to at-risk populations are also at risk for being marginalized or adopting passive roles when they are isolated in teams (*22, 25, 26*). Once they reach the third year, however, they are very likely to graduate. The focus should then shift to preparing them for the professional world where no one will be protecting them, and so this criterion may be dropped.

There are three principal ways to get the information needed to form teams using those rules:

1. On the first day of class, have the students fill out a survey containing several questions and an hour-by-hour matrix of the week, similar to the form at *<http://www.ncsu.edu/felder-public/CL_forms.doc>*. On the form, the students write their grades in selected prerequisite courses, times they are not available to meet outside class with their teams, and—if the criterion related to at-risk minorities is to be used—their gender and ethnicity. Use the surveys to form the groups, following the guidelines given above and using grades in prerequisite courses as the measure of ability.

2. Use *Team Maker*®, an on-line team-forming instrument developed at the Rose-Hulman Institute of Technology (*27*). The students enter the requested information into a database, the instructor specifies the sorting criteria, and Team Maker does the sorting. Sorting with Team Maker tends to be more reliable and much faster than manual sorting.

3. Let students self-select into groups, stipulating that no group may have more than one student who earned A's in one or two specified prerequisite courses. While not perfect, this system at least assures that the very best students in the class do not cluster together, leaving the weaker ones to fend for themselves.

Promoting positive interdependence

- *Assign different roles to team members (e.g. coordinator, recorder, checker, group process monitor), rotating the roles periodically or for each assignment.* The coordinator reminds team members of when and where they should meet and keeps everyone on task during team meetings; the recorder prepares the final solution to be turned in; the checker double-checks the solution before it is handed in and makes sure the assignment is turned in on time.; and the monitor checks to be sure everyone understands the solutions and the strategies used to get them. In teams of three, the coordinator may also assume the duties of the monitor.
- *Use Jigsaw to set up specialized expertise within each team.*
- *Give a bonus on tests (typically 2–3 points) to all members of teams with average test grades above (say) 80%.* The bonus should not be tied to each person on the team getting a certain grade, which would put too much pressure on weaker members of the team and make it impossible for teams with one very weak student to ever get the bonus. Linking the bonus to the team average grade gives all team members an incentive to get the highest grade they can and motivates the stronger students to tutor their teammates.
- *If an oral report is part of the team project, a short time before the report is given the instructor arbitrarily designates which team member should report on each part of the project.* Normally different team members take primary responsibility for different parts of the project and report on those parts, making it unnecessary for their teammates to understand what they did. When the proposed technique (which should be announced when the project is first assigned) is adopted, each student must make sure everyone on the team can report on what he or she did. This method provides both positive interdependence and individual accountability.

Providing individual accountability

- *Give individual tests that cover all of the material on the team assignments and projects.* Tests are frequently not given in traditional project-based courses such as laboratories and capstone research or design courses. Even if the tests only count for a relatively small portion of the course grade, their presence works against the familiar phenomenon of some team members doing little or none of the work and getting the same high course grades as their more responsible teammates.

- *In lecture courses (as opposed to project-based courses), include group homework grades in the determination of the final course grade only when a student has a passing average on the individual exams.* This policy—which should be announced in writing on the first day of class—is particularly important in required courses that are prerequisites for other courses in the core curriculum.

- *Make someone on the team (the process monitor) responsible for ensuring that everyone understands everything in the report or assignment that the team hands in.* The monitor should also make sure everyone participates in the team deliberations and that all ideas and questions are heard.

- *Make teams responsible for seeing that non-contributors don't get credit.* A policy that only contributors' names should go on assignments and reports should be announced at the beginning of the course, and reminders of the policy should be given to students complaining about hitchhikers on their teams. Most students are inclined to cut their teammates some slack initially, but if the the hitchhikers continue to miss meetings or fail to do what they were supposed to do, eventually the responsible team members get tired of being exploited and begin to implement this policy.

- *Use peer ratings to make individual adjustments to team assignment grades.* In a fairly simple but effective peer rating system, students rate one another on specified criteria for good team citizenship and the ratings are used to compute individual multipliers of the team grade that may range from 1.05 to 0 (*28*). An on-line system currently under development called CATME (Comprehensive Assessment of Team Member Effectiveness] computes a similar adjustment factor but also provides detailed feedback to team members on the skills and attitudes they need to work on and alerts the instructor to the existence of problematic situations (*29*). The ratings should be based primarily on responsible team behavior and not the percentages of the total effort contributed by each team member. Schemes of the latter sort move instruction away from the cooperative model toward individual competition, with a consequent loss in the learning benefits and skill development that cooperative learning promotes.

- *Provide last resort options of firing and quitting.* When a team has an uncooperative member and everything else has been tried and failed, the other team members may notify the hitchhiker in writing that he/she will be fired if cooperation is not forthcoming, sending a copy of the memo to the instructor. If there is no improvement after a week or if there is and the behavior later resumes, they may send a second memo (copy to the instructor) that he/she is no longer with the team. The fired student should meet with the instructor to discuss options. Similarly, students who are consistently doing all the work for their team may issue a warning memo that they will quit unless they start getting cooperation, and a second memo announcing their resignation from the team if the cooperation is not forthcoming. Students who get fired or quit must find a team of three willing to accept them, otherwise they get zeroes for the remaining assignments.

Help students develop teamwork skills

- *Establish team policies and expectations.* As part of the first assignment, have teams generate and sign a list of policies and expectations (e.g. being prepared before team sessions, calling if they have time conflicts, etc.). Have them sign the list and make copies for themselves and you. For an illustrative set of procedures, see <*www.ncsu.edu/felder-public/CL_forms.doc*>.
- *Keep groups intact for at least a month.* It takes at least that long for the teams to encounter problems, and learning to work through the problems is an important part of teamwork skill development.
- *Provide for periodic self-assessment of team functioning.* Every 2–4 weeks, have teams respond in writing to questions such as:

  How well are we meeting our goals and expectations?
  What are we doing well?
  What needs improvement?
  What (if anything) will we do differently next time?

- *Give students tools for managing conflict.* Caution them that dealing with conflicts quickly and rationally can avoid later serious problems that are almost certain to arise if they attempt to ignore the conflicts. Introduce them to *active listening*:

  – Students on one side of a dispute make their case without interruptions, then students on the other side have to repeat it to the initial group's satisfaction,

  – The second side then makes its case uninterrupted, and the first side has to repeat it to the second side's satisfaction.

  - The students then work out a solution. Once the students have articulated their opponents' cases, the solution frequently comes very easily.

The instructor should facilitate active listening sessions for groups in conflict, mainly making sure the rules of the procedure are followed.

- *Use crisis clinics to equip students to deal with difficult team members.* Two to three weeks after group work has begun, you will start hearing complaints about certain problematic team members, such as hitchhikers or dominant students who insist on doing the problems their way and discount everyone else's opinions. Use these characters as bases for ten-minute *crisis clinics* in class, in which the students brainstorm and then prioritize possible group responses to specified offending behaviors (*23*). At the end of this exercise, the teams leave armed with several excellent strategies for dealing with the problem, and the problem students in the class are on notice that their team members are likely to be ready for them in the future, which may induce them to change their ways.

General Suggestions

- *Start small and build.* If you've never used cooperative learning, consider starting with small group activities in class. See Felder and Brent (*30)* for suggestions about how to implement active learning effectively. Once you're comfortable with that, try a team project or assignment, and gradually build up to a level of cooperative learning with which you are comfortable.
- *At the start of the course, explain to students what you're doing, why you're doing it, and what's in it for them.* Let them know what they'll be doing in teams, what procedures you'll follow, and what your expectations are. Then tell them why you're doing it, perhaps noting that it will help prepare them for the type of environment most of them will experience as professionals, and sharing some of the research results (particularly those relating to higher grades). The section in this chapter on research support for cooperative learning provides useful material of this nature.
- *Make team assignments more challenging than traditional individual assignments.* CL works best for challenging problems and activities that require higher-level thinking skills. Students resent having to spend time in teams on assignments they could easily complete individually.
- *Don't curve course grades.* It should be theoretically possible for every student in a class to earn an A. If grades are curved, team members have

little incentive to help each other, while if an absolute grading system is used (e.g., a weighted average grade of 92–100 is guaranteed an A, 80–91 is guaranteed a B, etc.), there is a great incentive for cooperation.

- *Conduct a midterm assessment to find out how students feel about teamwork.* At about mid-semester, ask students to report anonymously on what's working and what's not working in their team. If many teams are having problems, spend some time in class on the relevant team skills. Most of the time, however, the assessment will show that most teams are working well. (Without the assessment, the instructor usually only hears the complaints.) If this is the case, announce the results at the next class session, so the few resistors become aware that they're in a small minority.
- *Expect initial resistance from students.* See Felder and Brent (*31*) for information on why the resistance occurs, what forms it is likely to take, and suggestions on how to deal with it.

Summary

Cooperative learning refers to work done by student teams producing a product of some sort (such as a set of problem solutions, a laboratory or project report, or the design of a product or a process), under conditions that satisfy five criteria: (1) positive interdependence, (2) individual accountability, (3) face-to-face interaction for at least part of the work, (4) appropriate use of interpersonal skills, and (5) regular self-assessment of team functioning. Extensive research has shown that relative to traditional individual and competitive modes of instruction, properly implemented cooperative learning leads to greater learning and superior development of communication and teamwork skills (e.g. leadership, project management, and conflict resolution skills). The technique has been used with considerable success in all scientific disciplines, including chemistry.

The benefits of cooperative learning are not automatic, however, and if imperfectly implemented, the method can create considerable difficulties for instructors, most notably dysfunctional teams and student resistance or hostility to group work. This paper offers a number of suggestions for forming teams, satisfying the five defining criteria of cooperative learning, and minimizing the problems. Instructors who have never used the approach are advised to move into it gradually rather than attempting a full-scale implementation on their first try, and to increase the level of implementation in subsequent course offerings. To an increasing extent, they should see the learning benefits promised by the research, and as their expertise and confidence in implementing the method continue to grow, student evaluations of the team experience should improve concurrently. Most importantly, instructors who are successful in using cooperative learning in their classes will have the satisfaction of knowing that they have significantly helped prepare their students for their professional careers.

Some years ago, one of us (RF) taught five chemical engineering courses in consecutive semesters to a cohort of students using cooperative learning (*32, 33*). The superiority of their performance and attitudes relative to a comparison group that was taught traditionally was consistent with the many other results reported on earlier in this chapter. Five years after most of the students had graduated they were surveyed and asked to reflect on what in their undergraduate college experience best prepared them for their post-graduation careers (*34*). Of the 50 respondents (out of 72 surveyed), 25 mentioned the problem-solving and time management skills they acquired by working on so many long and difficult assignments, 23 mentioned a variety of benefits gained from working in teams on homework, and no other feature of the curriculum got more than eight mentions. In their open comments, almost every respondent spoke positively about group work, mentioning its learning benefits and/or the interactions with classmates that it fostered. For example, "*I formed very close relationships with my group members that remain today. I realized that I wasn't alone in struggling with new concepts and could garner support and help from teammates.*" and "*Being forced to meet other students through required groupwork...kept me in the course long enough to develop the skills and self-confidence necessary to continue on in the CHE curriculum.*" No one said anything negative about group work, although two respondents indicated that they disliked it initially and only later came to see its benefits. We don't guarantee a retrospective evaluation this positive to everyone who uses cooperative learning, but we believe the possibility of it makes the effort worthwhile.

References

1. Johnson, D. W.; Johnson, R. T.; Stanne, M. E. *Cooperative Learning Methods: A meta-analysis.* University of Minnesota, Minneapolis: Cooperative Learning Center, 2000; http://www.co-operation.org/pages/cl-methods.html (accessed October 5, 2006).
2. Smith, K. A.; Sheppard, S. D.; Johnson, D. W.; Johnson, R. T. Pedagogies of engagement: Classroom-based practices. *J. Engr. Education* **2005**, *94*, 87-101.
3. Springer, L.; Stanne, M. E.; Donovan, S. *Effects of small-group learning on undergraduates in science, mathematics, engineering, and technology: A meta-analysis.* National Institute for Science Education: Madison, WI, 1997; http://www.wcer.wisc.edu/nise/CL1/CL/resource/R2.htm (accessed October 5, 2006).

4. Terenzini, P. T.; Cabrera, A. F.; Colbeck, C. L.; Parente, J. M.; Bjorklund, S. A. Collaborative learning vs. lecture/discussion: Students' reported learning gains. *J. Engr. Education* **2001**, *90*, 123-130.
5. *How People Learn: Brain, Mind, Experience, and School;* Bransford, J.; Brown, A. L.; Cocking, R. R., Eds.; National Academy: Washington, DC, 2000; http://www.nap.edu/books/0309070368/html/ (accessed October 6, 2006).
6. McKeachie, W. J. *Teaching Tips: Strategies, Research, and Theory for College and University Teachers* (11th ed.); Houghton Mifflin: Boston, MA, 2002.
7. Johnson, D. W.; Johnson, R. T.; Smith, K. A. *Active Learning: Cooperation in the College Classroom*, (2nd ed.); Interaction Book: Edina, MN. 1998.
8. *Peer-led Team Learning: A Guidebook*; Gosser, D. K.; Cracolice, M. S.; Kampmeier, J. A.; Roth, V.; Strozak, V. S.; Varma-Nelson, P. Eds.; Prentice-Hall: Upper Saddle River, NJ, 2001.
9. Sarquis, J. L.; Dixon, L. J.; Gosser, D. K.; Kampmeier, J. A.; Roth, V.; Strozak, V. S.; Varma-Nelson, P. The Workshop Project: Peer-led team learning in chemistry. In *Student-assisted Teaching: A Guide to Faculty-student Teamwork*; Miller J. E.; Groccia J. E.; Miller M. Eds.; Anker: Bolton, MA, 2001; pp 150–155.
10. Tien, L. T.; Roth, V.; Kampmeier, J. A., Implementation of a Peer-Led Team Learning Approach in an Undergraduate Organic Chemistry Course. *J. Research in Sci. Teaching* **2002**, *39*, 606-632.
11. Nurrenbern, S. C.; Robinson, W. R. Cooperative Learning: A Bibbliography. *J. Chem. Educ.* **1997**, *74*, 623-624.
12. Hinde, R. J.; Kovac, J. Student active learning methods in physical chemistry. *J. Chem. Educ.* **2001**, *78*, 93-99.
13. Wamser, C. C. Peer-Led Team Learning in Organic Chemistry: Effects on Student Performance, Success, and Persistence in the Course. *J. Chem. Educ.* **2006**, *83*, 1562-1566.
14. Pederson, P. An alternative biological sciences laboratory. *J. Coll. Sci. Teaching* **1994**, *24*, 162-165.
15. Dougherty, R. C. Grade/study-performance contracts, enhanced communication, cooperative learning, and student performance. *J. Chem. Educ.* **1997**, *74*, 722-726.
16. Hanson, D.; Wolfskill, T. Process workshops: A new model for instruction. *J. Chem. Educ.* **2000**, *77*, 120-130.
17. Wolfskill, T.; Hanson, D., LUCID: A New Model for Computer-Assisted Learning. *J. Chem. Educ.* **2001**, *78*, 1417-1424.
18. Bowen, C. W. A Quantitative Literature Review of Cooperative Learning Effects on High School and College Chemistry Achievement. *J. Chem. Educ.* **2000**, *77*, 116-119.

19. Hake, R. R. Interactive-Engagement vs. Traditional Methods: A Six-Thousand-Student Survey of Mechanics Test Data for Introductory Physics Courses. *Am. J. Phys.* **1998**, *66*, 64-74.
20. Cabrera, A. F.; Crissman, J. L.; Bernal, E. M.; Nora, A.; Terenzini, P. T.; Pascarella, E. T. Collaborative Learning: Its Impact on College Students' Development and Diversity. *J. Coll. Stud. Development* **2002**, *43*, 20-34.
21. Towns, M. H.; Kreke, K.; Fields, A. An Action Research Project: Student Perspectives on Small-Group Learning in Chemistry. *J. Chem. Educ.* **2000**, *77*, 111-115.
22. Felder, R. M.; Brent, R. Cooperative Learning in Technical Courses: Procedures, Pitfalls, and Payoffs, ERIC Document Reproduction Service, ED 377038, 1994, http://www.ncsu.edu/felder-public/Papers/Coopreport.html (accessed October 6, 2006).
23. Felder, R. M.; Brent, R. Effective Strategies for Cooperative Learning. *J. Cooperation & Collaboration in College Teaching* **2001**, *10*, 69–75, http://www.ncsu.edu/felder-public/Papers/CLStrategies(JCCCT).pdf (accessed October 6, 2006).
24. Oakley, B.; Felder, R. M.; Brent, R.; and Elhajj, I. Turning Student Groups into Effective Teams. *J. Student Centered Learning* **2004**, *2*, 9–34, *http://www.ncsu.edu/felder-public/Papers/Oakley-paper(JSCL).pdf* (accessed October 6, 2006).
25. Heller, P.; Hollabaugh, M. Teaching Problem Solving through Cooperative Grouping. Part II: Designing Problems and Structuring Groups. *Am. J. Phys.* **1992**, *60*, 637-644.
26. Rosser, S. V. Group Work in Science, Engineering, and Mathematics: Consequences of Ignoring Gender and Race. *Coll. Teaching* **1998**, *46*(3), 82-88.
27. Cavanaugh, R.; Ellis, M.; Layton, R.; Ardis, M. Automating the Process of Assigning Students to Cooperative-Learning Teams, *Proceedings of the 2004 ASEE Annual Conference*, Salt Lake City, UT, 2004. See *http://teammaker.cs.rose-hulman.edu/* (accessed October 6, 2006) for information about obtaining and using this instrument.
28. Kaufman, D. B.; Felder, R. M.; Fuller, H. Accounting for Individual Effort in Cooperative Learning Teams. *J. Engr. Education* **2000**, *89*, 133-140, http://www.ncsu.edu/felder-public/Papers/Kaufmanpap.pdf (accessed October 6, 2006).
29. Ohland, M. W.; Loughry, M. L.; Carter, R. L.; Bullard, L. G.; Felder, R. M.; Finelli, C. J.; Layton, R. A.; Schmucker, D. G. The Comprehensive Assessment of Team Member Effectiveness (CATME): A New Peer Evaluation Instrument. *Proceedings of the 2006 ASEE Annual Conference*, Chicago, IL, June 2006. Information about CATME may be found at *http://www.catme.org* (accessed October 6, 2006).

30. Felder, R. M.; Brent, R. Learning by Doing. *Chem. Engr. Education* **2003**, *37*, 282–283, *http://www.ncsu.edu/felder-public/Columns/Active.pdf* (accessed October 6, 2006).
31. Felder, R. M.; Brent, R. Navigating The Bumpy Road to Student–Centered Instruction. *College Teaching* **1996**, *44*, 43-47, *http://www.ncsu.edu/felder-public/Papers/Resist.html* (accessed October 6, 2006).
32. Felder, R. M. A Longitudinal Study of Engineering Student Performance and Retention. IV. Instructional Methods and Student Responses to Them. *J. Engr. Education* **1995**, *84*, 361-367, *http://www.ncsu.edu/felder-public/Papers/long4.html* (accessed October 6, 2006).
33 Felder, R. M.; Felder, G. N.; Dietz, E. J. A Longitudinal Study of Engineering Student Performance and Retention. V. Comparisons with Traditionally-Taught Students. *J. Engr. Education* **1998**, *87*, 469-480, *http://www.ncsu.edu/felder-public/Papers/long5.html* (accessed October 6, 2006).
34. Felder, R. M. The Alumni Speak. *Chem. Engr. Education* **2000**, *34*, 238-239, *http://www.ncsu.edu/felder-public/Columns/alumni.html* (accessed October 6, 2006).

Chapter 5

Collaborative and Project-Based Learning in Analytical Chemistry

Thomas J. Wenzel

Department of Chemistry, Bates College, Lewiston, ME 04240

Collaborative and project-based learning are used to enhance the learning outcomes of undergraduate analytical chemistry courses. The learning outcomes that can be realized through the use of these methods are described. The advantages of these methods in promoting student learning of content, techniques, and other skills such as communication, team work, leadership, and problem solving is discussed. The format that is used in the class and laboratory is described, and examples of the projects that students undertake in the lab are presented.

The undergraduate analytical chemistry curriculum is often structured into two courses, one generally referred to as quantitative analysis and the other as instrumental analysis. The history of this course structure has been summarized, and both textbooks and changes in the discipline were important in shaping the current situation (*1*). The textbooks available today for undergraduate analytical chemistry courses generally reinforce the quantitative and instrumental structure. The distinction between quantitative and instrumental methods is a false one as instruments are used to perform quantitative analyses and every quantitative analysis course most likely includes some aspects of instrumentation. A further reality is that the textbooks within these two categories primarily consist of chapters distinguished by the type of analysis method. These books include far more analysis methods than can be reasonably covered in a one-semester course, leaving the instructor with difficult choices to make about what material and methods ought to be covered in the limited amount of available time.

One of the important steps in designing a course is to identify a set of learning goals and objectives. One useful method for identifying learning goals is to use a device known as the Total Goals Inventory survey (*2,3*). A question to consider is the extent to which the textbooks that are available for the quantitative and instrumental analysis courses shape the typical goals identified by most instructors of analytical chemistry courses. If asked to specify a set of learning goals, I suspect that many instructors would provide a list of analysis methods that they intend to cover in the course, something reinforced by published surveys of the topics included in undergraduate analytical chemistry courses (*1,4*). Instructors might speak of the importance of other learning goals, perhaps chief among them having the students learn problem-solving skills, but it is also worth interrogating what is meant by "problem-solving." Is it the quantitative problems that are included at the end of each chapter of textbooks or is it the type of problem-solving used by practicing analytical chemists in industrial and other careers (*5*)? My guess is that, more often than not, problem-solving in analytical chemistry courses mostly involves solving problems at the end of each chapter.

A further consideration is the structure of the laboratory component of the quantitative and instrumental analysis course. A common tendency is to view the laboratory as a way of reinforcing material covered in the class, so that students in the quantitative and instrumental analysis laboratory conduct measurements using many of the methods included in the syllabus. Furthermore, these experiences are often set up as a series of one- or two-week experiments that are done by the entire class or rotated through depending on the availability of equipment. The goal, as best as possible, is to introduce the students to the array of equipment available at the institution and as many analysis methods as possible. More often than not, I suspect that students are given a set of unknowns where the instructor knows the amount of analyte in the sample and the student is expected to determine it.

The ultimate result is that analytical chemistry courses are often designed with a content and coverage structure. A set of topics is identified that the instructor intends to cover, and the class and laboratory are used to develop these topics.

Learning Outcomes for an Undergraduate Analytical Chemistry Course

A key question to consider is whether such an approach does justice to the field of analytical chemistry, and whether such an approach encompasses the learning goals that ought to be included within an undergraduate analytical chemistry course.

One way to consider this question is to ask what employers of our students look for as desirable skills. In one recommendation for a student, I was asked to evaluate the student's intellectual curiosity and imagination, written and oral communication skills, initiative, sensitivity to the interests and views of others, ability to take directions, ability to cope with ambiguity, positive interaction with others, common sense and good judgment, capacity to follow through, sense of humor, and adaptabilty and flexibility. Others have discussed this as well by surveying what skills people in their careers found most important (*5,6*). Problem-solving, interpersonal, technical writing, management and communication skills were among the most highly rated. The ability to arrive at informed judgment, to function in a global community, and the ability to work with others, especially in teams, were also highly valued for success in a career. The traditional structure of analytical chemistry courses and laboratories do not emphasize the development of these skills.

Another way to consider whether analytical chemistry courses do justice to the field is to examine the steps involved in the analytical method. One rendition of the analytical method that I find useful was put forward by Laitenen and Harris in 1975 (*7*). They provide five steps in the analytical procedure.

- Definition of the goal
- Sampling
- Separating the sought-for constituent from other species present in the sample
- Measurement of the desired substance
- Evaluation and interpretation of the data

The approach that I first used in the laboratories with my courses emphasized only the last two steps in this process. Students were given unknowns that were soluble in the desired solvent or easily converted to a completely soluble form amenable for analysis, and then measured, evaluated, and interpreted the data. The steps of actually defining the goal, conducting real sampling, and considering matrix effects and determining how to separate the sought-for constituent from other species were not explored in a meaningful way.

I eventually came to realize that I wanted my courses to have a much broader set of learning goals. A set of learning outcomes that I have found to be particularly helpful are provided below (*8*). These outcomes are grouped into four categories: knowledge outcomes, skills outcomes, affective outcomes, and learned abilities.

- *Knowledge outcomes* – involve particular areas of disciplinary or professional content that students can recall, relate, and appropriately deploy.

- *Skills outcomes* – involve the learned capacity to do something – for example, think critically, communicate effectively, productively collaborate, or perform particular technical procedures – as either an end in itself or as a prerequisite for further development.
- *Affective outcomes* – involve changes in beliefs or in the development of particular values, for example, empathy, ethical behavior, self-respect, or respect for others.
- *Learned abilities* – involve the integration of knowledge, skills, and attitudes in complex ways that require multiple elements of learning. Examples embrace leadership, teamwork, effective problem-solving, and reflective practice.

I would contend that the traditional way we have taught analytical chemistry does an excellent job with *knowledge outcomes* and includes the "performing particular technical procedures" aspects of *skills outcomes*. I would argue that traditional science courses do not really teach students the *skills outcomes* of thinking critically, communicating effectively, or productively collaborating, and essentially omit any emphasis on *affective outcomes* and *learned abilities*. Some of these areas may develop in students inadvertently as they participate in courses, but the courses are often not specifically designed in a way that provides opportunities for students to grow in these areas. These learning outcomes, which are far more encompassing that those I had when I first started teaching, indicate that there is the opportunity for us to accomplish much more in our undergraduate courses. It is important to note that the use of instructional methods that expand the learning outcomes does not mean that disciplinary content in the course must be sacrificed (*9*). The exercises and problems that are used must still be based on content areas of analytical chemistry, so that appropriate content is introduced in a different context.

Format of my Courses

As a way of removing the arbitrary distinction between quantitative and instrumental analysis, and of better emphasizing the importance of separation processes within the area of chemical analysis, I have altered the material in my courses and now teach *Separation Science* and *Analytical Spectroscopy and Electrochemistry* (*10,11*). More important than the adjustment of the topics included in these courses is the change in the manner in which the classroom and laboratory are structured. My classroom environment is now organized to promote collaborative learning. Students are encouraged to cooperate both in and out of class on the course work. My laboratory is structured so that the students work in small groups to undertake semester-long projects.

Collaborative Learning in the Classroom

I started collaborative learning with my students before realizing that there was a rich literature available on its effectiveness. I began having my students collaborate in class because I was discouraged with the performance of some students on my exams on chemical equilibrium. Even though all the students seemed to understand the concepts during my lectures, answers to homework assignments and on exams indicated that several often did not. I became concerned that the lecture format was too passive a form of learning and wanted to more actively engage my students in the learning process. Studies show that lecturing is an ineffective way to promote learning with many individuals (*12,13,14,15*). Many people in lectures have difficulty staying on task and the lecturer has almost no sense of whether the students actually comprehend the material. Conversely, cooperative learning, a strategy in which in which students work together on questions and problems that are provided in the class, has been extensively studied (over 400 carefully constructed studies at the college level), with the following demonstrated outcomes resulting from these assessments (*13*).

- Statistically significant improvements in academic achievement
- Better reasoning and critical thinking skills
- Proposed more new ideas when presented with problems
- Transferred more of what was learned in prior situations to new problems
- Reduced levels of stress
- Promotes more positive attitudes toward subject and instructional experience – faculty get to know students better
- Decreased absenteeism – whereas I used to have about 10-15 students out for each lecture in an introductory class of 60 students, now with cooperative learning its rare if I even have one student out, and that's even when my class meets at 8:00 am.
- Improved student commitment
- Greater motivation toward learning
- Better student retention (especially for women and minorities). This occurs because retention is improved if students are socially and academically involved, and cooperative learning has the ability to foster both of these.

While my methods do not rigorously adhere to the criteria set out for a true "cooperative learning" experience (*13*), I have observed that the collaborative methods I use produce all of the above outcomes in my own courses.

Many possible ways to structure cooperative learning groups and to utilize cooperative learning methods have been described in the literature (*13,16,17,18,19,20)*). Examples of the use of cooperative or collaborative group learning in general *(21-31)*, analytical *(32-44)*, biological *(45,46)*, inorganic

(47), organic *(48-53)*, and physical *(54-56)* chemistry courses have been described, although this is by no means an exhaustive list of all of the published examples on the use of group learning methods in the laboratory or classroom components of chemistry courses.

In my classes, I gather information about the students on the first day of class and then use that to assign groups that work together for the remainder of the term. These groups have three or four students in them, and I try to make them as heterogeneous as possible by gender, race, ethnicity, year of study, etc. Although, after regularly noticing at evening problem-solving sessions that I often organize that the students of color worked together irrespective of whether they were in same group or not, I have now begun to put them together in class groups. I prepared the in-class exercises from my previous lecture notes. Questions that I posed to the class were incorporated into the in-class exercises. Examples of calculations that I performed at the board (e.g., calculating the pH of a solution of a weak base) were put onto the exercises. These exercises encompass quantitative and conceptual (e.g., factors contributing to peak broadening in chromatographic separations) topics. I spend a considerable amount of time during the first day of class going over the format for the collaborative learning and carefully explaining my expectations of the students. During the semester I occasionally remind them of these expectations.

An in-class exercise is distributed to the students and the groups begin to work on the first question. They are not allowed to use a textbook, although any appropriate tables of data (e.g., a table of equilibrium constants) are provided. I circulate among the groups to facilitate their discussion. If I hear a point of confusion about the question that other groups may have as well, I get everyone's attention and clarify what is being asked. Groups then go back to work. If an individual within a group is saying something or has written down something that is especially pertinent to the question, I indicate that the group should reflect upon that comment. If a group is stuck or heading in the wrong direction, then I ask leading questions rather than providing direct answers. Sometimes when limited discussion is taking place among the groups, I have found that it helps to leave the room. Invariably, when I return, the groups are fully engaged in the question. It is essential to remain patient and to give the students the time and space to work on the question. If the problem has key points of recognition along the way, when every group has appreciated a particular point I get their attention and spend a brief period of time summarizing the concept that has just been realized. Then the groups go back to the remainder of the problem.

Positive Attributes I Observe with Collaborative Learning

There are many attributes of this method that I find especially worthwhile, in addition to those listed above. One is that there are now many instructional resources besides myself. Students who understand a concept are expected to explain it to other members of the group who do not. This is beneficial to both the student doing the explaining and the student receiving it. The classroom atmosphere is far less formal than lectures with considerable give-and-take between me and the students. I appreciate what actually confuses them about a problem such that I can provide better guidance in helping them understand the concept. The students are also better at identifying what they do and do not comprehend about the problem. In my larger general chemistry class with 60 students, the use of collaborative learning helps me learn more of the student's names and makes it easy to take attendance (*57,58*). Before instituting cooperative learning, I often had 5-15 of the 60 students missing each class. Since I started cooperative learning, it is rare to have more than a single student out, even when I have taught the class at 8 am.

I teach at a college with a residential campus and have the ability to "require" the groups to meet outside of class. Each student must submit an individual homework assignment, but it is my expectation that all members of a group have the same answers. I made a purposeful decision to have each student submit homework answers for several reasons. One is that many of the students first like to try the homework on their own and to then get together to share their work. Another is that students often like to work with others besides those in their assigned groups and I want to encourage any form of cooperation. Finally, each student has a completed set of homework answers in her or his own writing, which I believe helps them study for the exams. What I have found is that the groups do get together outside of class to share homework answers or work together on the homework. To further facilitate cooperation outside of class, I reserve a room on campus the evening before the homework assignment is due. Students can then come and work in any arrangement they choose. I am there as a facilitator only and not to work the problems at the board. By encouraging the students to cooperate out of class, I have found that they spend much more time working on course material. Essentially every student submits every homework assignment with all of the questions worked out. This network of support pays off as grades on my exams, which are administered individually, went up significantly once I instituted cooperative learning (*57*).

For example, before instituting collaborative learning in my introductory chemistry class, less than 15% of the students got a grade of 95% or higher on a quiz on the stoichiometry unit. Since instituting collaborative learning, about 60% of the students have gotten a grade of 95% or higher. On the final for a special topics upper-level course that involves the interpretation of NMR spectra, fewer than 10% of the students got all ten of the problems correct before I began using collaborative learning. Since instituting collaborative learning, 45% of the students who have taken the course correctly answered all ten of the problems. Before instituting collaborative learning in the unit on chemical equilibrium, it was common to have a few students in the class make essentially no progress in understanding the complex, simultaneous equilibria problems covered at the end of the unit. These students would often score in the single digits on the 100-point exam on this topic. As a result, class averages in the 45% range were common. Since instituting cooperative learning, single-digit scores no longer occur and class averages are typically in the 65% range for exams on these exceedingly complex problems.

Avoiding Dysfunctional Groups

When using group work in the class or lab, the instructor must always be sensitive to the possibility of dysfunctional groups. Reminding the students of their responsibilities within group learning is one strategy to counter this possibility. I am also attentive to the work of the groups and if I sense I problem, intervene either by talking to an individual who is not functioning as a member of a team or by meeting with the entire group if I sense that the group is not working well together. I also employ a peer- and self-assessment process in both a formative manner part way through the course and in a summative manner at the end of the course. A review of peer- and self-assessment practices that includes my own methods and observations is available for those who are interested (*59*). The formative assessment is used to identify whether any of the groups are having problems. The summative assessment is used in context with my own observations to award each student appropriate credit for their functioning within the groups. A portion of the class (~5%) and laboratory grade (20%) is awarded for participation, which encompasses both effort and the student's effectiveness at working within a group. Despite the occasional group that needs intervention, the utilization of collaborative learning in my courses has been an unqualified success. In fact, I was so satisfied with the results of collaborative learning in my smaller analytical courses (12-22 students) that I now use it in my introductory chemistry course which has 60 students in it (*60*).

Project-based Laboratory Experiments

The laboratory component of my separation science course is structured such that the students work in groups of two or three and undertake a single analysis project for the entire semester. I opted for a single project because I wanted my students to become fully engaged in an ambitious undertaking that they would carry through from start to finish. I assign the groups and try to make them as different as possible from the class groups. Examples of the types of projects that have been done in the course are listed.

- Determination of benzene and toluene in air using GC-MS
- Determination of trihalomethanes in drinking water using GC-MS

- Determination of anions in soil using ion chromatography
- Determination of nitrate and nitrite in cured meats using ion chromatography
- Determination of the amino acid content in milk, popcorn and beer using reversed-phase LC with fluorescence detection
- Determination of caffeine, theophylline, and theobromine levels in chocolate using reversed-phase LC with ultraviolet detection
- Determination of polycyclic aromatic hydrocarbons in charbroiled meats, oysters, or creosote using GC-MS or LC with fluorescence detection
- Determination of heavy metals in waste water treatment plant sludges using ICP-AES
- DNA restriction fragment analysis using capillary electrophoresis

Students are only provided with the title of the project and, since this is the first course they have ever had in analytical chemistry and they are not ready yet to compare different measurement techniques, the particular instrumental method that will be used to conduct the analysis. Using *Scifinder Scholar*, each group must first identify and gather suitable literature that can be used to design the specific procedures that will be used to carry out the project. I meet individually with each group to teach them the basics of operating their equipment and to identify particular aspects that they should look for when reviewing the literature articles. By the fifth week of the course each group must turn in a single proposal that discusses the literature and specifies and defends the entire procedure that will be used to sample and carry out the analysis. After being approved, the students then undertake the project.

Once checked out on the instrument, and clear about any potential safety hazards with aspects of the sample workup, the students are able to use the laboratory at off-hours. Each student must keep a log of her or his hours and is required to spend a minimum of 30 hours over the semester on the project. The hours include the time spent reviewing background literature, preparing the proposal, sampling, and performing any laboratory activities, but does not include the time that is spent preparing the final oral and written reports. Usually I find that several members of the class work 60 or more hours on the project. Students take advantage of the opportunity to work off-hours, and after clearing it with me, often spend time in the evening or weekends in the lab. This can be especially helpful if a group wishes to spend an entire day on the instrument running standards and samples. Over the years, groups that have conducted analyses of air samples for volatile organic compounds have routinely gotten up at 6 am to sample air that should likely have reduced levels of chemicals from automobile exhaust to compare to samples gathered later in the day. Other groups have sampled water from high alpine lakes in the White Mountains of New Hampshire (but not counted the time hiking in to get these samples within their 30 hours). For most students, the open lab format is the first chance they have to appreciate that science does not occur in only three-hour blocks of time. The extra hours they put in occur voluntarily as the students become engaged in the work and want to make more progress on the project.

On the last day of lab, each group gives an oral presentation on their project to the rest of the class. Each student is required to speak for about ten minutes on some aspect of the project (groups usually divide this into an introduction, the experimental details, and the results and discussion). The oral report has provided a significant incentive for students to work hard on their project, as they do not want to be embarrassed in front of their peers by having a lack of data. Each student must also submit an individually written final report that takes the form of a journal article in *Analytical Chemistry*. Group members are encouraged to discuss their method, data, and conclusions. I also indicate that each report may have certain figures or tables that are identical since it is acceptable for them to decide as a group how to present a standard curve, a series of chromatograms, or a table of measured values for a set of unknowns. But they are also informed that, when it comes to writing the actual text to describe the information in the figures and tables and to describe their conclusions, it must be their own words.

The grade for the lab is based on their participation in the project (which is assessed by the hours they have spent on the project, my observations of their work in lab, and the self- and peer assessment), their oral presentation (weighted toward the effort that went into developing it), and their written report.

Advantages of Laboratory Projects

I have found many advantages to having students undertake semester-long projects when compared to the one or two-week experiments I used in my first ten years of teaching. The projects incorporate the entirety of the analytical procedure, since students must define aspects of the problem, devise a procedure for sampling and sample workup, collect samples, workup samples with complex matrices into a form suitable for analysis, perform measurements, and interpret data. Every group finds reports in the literature that present different ways of possibly collecting or working up the samples. These must be judged using criteria such as reliability, ease of implementation, expense, and the time involved in deciding the "best" method for the analysis. Every group encounters unanticipated problems in implementing their project. Some find that the method included in their proposal presents intractable problems and must be changed during the term. Others find that the procedures described in the literature do not work as simply as stated and must be modified. Many groups find that the sample workup process is more complicated than their initial impressions based on the descriptions in the literature, and takes longer than they initially expected. By the end of the semester, with the benefit of hindsight, they realize how much faster they could now carry out the project were they to do it again. Each of the projects involves aspects of experimental design. Students must ask and answer questions during the execution of the project.

The students also get experience at working in a team. I stress the importance of having them coordinate tasks and responsibilities in an effort to make as much progress as possible during the limited amount of time available during the semester. Most groups get into a coordinated routine by the midpoint of the semester such that one individual might come in early in the day to start a sample workup and another comes in later either to finish it up or run samples on the instrument. I am insistent that the students not adopt specific roles (e.g., one person only does sample workup while another only runs the instrument) but must participate in and understand all aspects of the project. I carefully monitor this situation through my observations in the lab and through the peer- and self-assessment.

The most gratifying outcome of the projects has been the degree to which the students are empowered by the experience. In my former laboratory format, when I had groups rotating through shorter experiments with each group working on a different instrument often spread out through different laboratories, it seemed like every time I entered a room the group was sitting and waiting for me because they had a question. It might have involved an apparent malfunction with the equipment or some uncertainty about what exactly the experiment asked them to do. Since switching to the projects, I have had many occasions where an instrument malfunctioned and the students used the printed or electronic manuals to begin troubleshooting the problem. When I come across a group in such a situation, I have them describe the problem and their diagnosis of what is wrong with the equipment, and the way in which they intend to fix it. If everything seems reasonable to me, I let them proceed with the repair.

A more common occurrence is that a question comes up related to a step in the sample workup, and I am not immediately available to answer it. Rather than trying to find me, the groups typically use their best judgment to make a decision. Sometimes they are right. Other times they are wrong. But the students are praised for their willingness to make a decision, and I use any mistakes as an opportunity for discussion and learning (*61*). In fact, projects sometimes fail or, more commonly, have situations in which a failure occurs along the way. When this occurs, the students either have to repeat the procedure or identify an alternate procedure, usually based on further review of the literature. Sometimes the students do not get a project to work at all. One example involved a group that tried to use capillary electrophoresis to perform DNA restriction fragment analysis. The group could never get reproducible results. But in the process of trying to get reproducible results, they consulted with technical staff at the company that manufactured the equipment, tested different columns and reagents, and did the other steps that one would carry out when trying to get a particular method to work. I made sure that they appreciated the beneficial learning experience they gained through the project, even when they were not able to get reproducible results.

The reality is that no group has ever truly completed their project, as they have never had enough time to analyze enough samples and validate the results. Even so, the students do appreciate each of the steps of the analytical process. They also realize how difficult it can be to obtain a reliable analytical number, especially when performing trace analysis.

One potential concern with my approach is that students only get experience with a single instrument rather than being introduced to many instruments. I have several responses to this issue. One is whether it is actually better to gain a very thorough knowledge of one sophisticated instrument, as students do get through the project, rather than the superficial introduction to equipment they used to get in my courses through a one or two-week experience. What I have observed is that the students, having learned the operation of a sophisticated piece of equipment that was at first intimidating, are then quite comfortable learning how to operate a new instrument in another context. Students in my department are introduced to other pieces of sophisticated instrumentation in labs in physical chemistry, advanced organic synthesis, and their required research projects, so over the entire curriculum they gain experience with many pieces of equipment. One last fact to consider is that no chemistry department will ever have all of the instruments an individual may encounter in a career. What is really important is having confidence in one's ability to learn the operation of a new piece of equipment. This confidence is what students should get from our analytical chemistry courses.

A reality is that the laboratory projects require more of my time, especially at the start of the term. I have had as many as eight projects occurring simultaneously, and getting all of the groups checked out on the equipment and through the early stages of identifying questions that they will need to address through their literature review is time consuming. The second half of the semester is noticeably different as the groups work quite independently and only come to see me when a significant problem or question arises. The situation over the second half of the term involves an interaction more like what occurs with my research students. The enhanced learning that occurs and the shear enjoyment of engaging with my students in a real investigation on something they actually care about easily justifies the extra expenditure of time required with the projects.

Summary

The use of collaborative learning and semester-long group projects has allowed me to incorporate into my courses many more of the learning outcomes described earlier. I still decide what areas of disciplinary content I want to include in my courses and design collaborative learning activities to incorporate such knowledge outcomes in my students. The use of cooperative exercises in the class and the decision making that accompanies the projects in the lab facilitates critical thought, communication at a variety of levels, and collaboration among my students. Self-respect and respect for others has the opportunity to develop through the use of group activities in which students must depend on each other when completing in-class exercises and the lab project. The group activities provide the chance for students to exhibit leadership and to develop teamwork skills. The projects require effective problem solving and offer the chance to engage in reflective practice. I firmly believe that the approach I use now provides my students a far greater skill set that will serve them well in whatever career path they choose after completing their undergraduate study.

There is a quote I especially like from *Walden; or, Life in the Woods* by Henry David Thoreau (*62*). This is a book my students and I read in a first-year seminar in which we examine the human relationship to land. Thoreau rails against lots of things in *Walden*, one of which is the way we educate college students. As Thoreau writes:

> "Which would have advanced the most at the end of a month – the boy who had made his own jackknife from the ore which he had dug and smelted, reading as much as would be necessary for this, - or the boy who had attended the lectures on metallurgy at the Institute in the mean while, and had received a penknife from his father?"

I think that Thoreau provides an excellent example of what type of experience provides a better education in the sciences, and I think it is incumbent on those of us who teach in the sciences to provide similar experiences for our students.

References

1. Christian, G. D. *Anal. Chem.* **1995**, *67*, 532A-538A.
2. Wenzel, T. J., *Anal. Chem.* **1995**, *72*, 659A-660A.
3. Angelo, T.A.,; Cross, K. P. *Classroom Assessment Techniques: A Handbook for College Teachers*; Jossey-Bass; San Francisco, CA **1993**.
4. Girard, J. E.; Diamant, C. T. *J. Chem. Educ.* **2000**, *77*, 646-648.
5. Thorpe, T. M.; Ullman, A. H. *Anal. Chem.* **1996**, *73*, 477A-480A.
6. Duch B. J.; Groh, S. E.; Allen, D. E. *The Power of Problem-Based Learning*; Stylus Publishing; Sterline, VA **2001**.
7. Laitenen, H. A.; Harris, W. E. Chemical Analysis, 2nd ed.; McGraw-Hill: New York, **1975**.

8. Ewell, P. T. *Accreditation and Student Learning Outcomes: A proposed Point of Departure*, September **2001**, Council for Higher Education Accreditation (CHEA) Occasional Paper, Washington, DC.
9. Wenzel, T. J. *Anal. Chem.* **1999**, *71*, 693A-695A.
10. Wenzel, T. J. *Anal. Chem.* **1995**, *67*, 470A-475A.
11. Wenzel, T. J. *Microchim. Acta* **2003**, *142*, 161-166.
12. Wenzel, T. J. *Anal. Chem.* **1999**, *71*, 817A-819A.
13. Johnson, D. W.; Johnson, R. T.; Smith, K. A. *Cooperative Learning: Increasing College Faculty Instructional Productivity* **1991**, ASHE-ERIC Higher Education Report No 4. Washington, D.C.: The George Washington University, Graduate School of Education and Human Development.
14. Penner, J. G. *Why Many College Teachers Cannot Lecture: How to Avoid Communication Breakdown in the Classroom* **1984**, Charles C. Thomas, Springfield, IL.
15. Bligh, D. A. *What's The Use of Lectures?* **1972**, Penguin, Harmondsworth, Eng.
16. Johnson, D. W.; Johnson, R. T. *Cooperation and Competition: Theory and Research* **1989**, Interaction Book Company, Edina, MN.
17. Blosser, P. E. *The Science Outlook: Using Cooperative Learning in Science Education* **1993**, ERIC-CSMEE, Columbus, OH, 1-9.
18. Wenzel, T. J. *Anal. Chem.* **1998**, *70*, 790A-795A.
19. Wenzel, T. J. *Anal. Chem.* **2000**, *72*, 293A-296A.
20. Wenzel, T. J. *Anal. Chem.* **2000**, *72*, 359A-361A.
21. Selco, J. I.; Roberts, Jr., J. L.; Wacks, D. B. *J. Chem. Educ.* **2003**, *80*, 54-57.
22. Shibley, Jr., I. A.; Zimmaro, D. M. *J. Chem. Educ.* **2002**, *79*, 745-748.
23. Oliver-Hoyo, M. T.; Allen, D; Hunt, W. F.; Hutson, J.; Pitts, A. *J. Chem. Educ.* **2004**, *81*, 441-448.
24. Wimpfheimer, T. *J. Chem. Educ.* **2004**, *81*, 1775-1776.
25. Kovac, J. *J. Chem. Educ.* **1999**, *76*, 120-124.
26. Farrell, J. J.; Moog, R. S.; Spencer, J. N. *J. Chem. Educ.* **1999**, *76*, 570-574.
27. Hanson, D.; Wolfskill, T. *J. Chem. Educ.* **2000**, *77*, 120-130.
28. Wright, J. C.; Millar, S. B.; Kosciuk, S. A.; Penberthy, D. L.; Williams, P. H.; Wampold, B. E. *J. Chem. Educ.* **1998**, *75*, 986-992.
29. Juhl, L. *J. Chem. Educ.* **1996**, *73*, 72-77.
30. Mills, P.; Sweeney, W. V.; Marino, R.; Clarkson, S. *J. Chem. Educ.* **2000**, *77*, 1161-1165.
31. Wenzel, T. J. *Gender, Science and the Undergraduate Curriculum: Building Two Way Streets*, Association of American Colleges and Universities, **2001**, 29-46.
32. Williamsson, V. M.; Rowe, M. W. *J. Chem. Educ.* **2002**, *79*, 1131-1134.
33. Ram, P. *J. Chem. Educ.* **1999**, *76*, 1122-1126.
34. Allison, J. *J. Chem. Educ.* **2001**, *78*, 965-969.
35. Houghton, T. P.; Kalivas, J. H. *J. Chem. Educ.* **2000**, *77*, 1314-1318.

36. Jackson, P. T.; Walters, J. P. *J. Chem. Educ.* **2000**, *77*, 1019-1025.
37. Wright, J. C. *J. Chem. Educ.* **1996**, *73*, 827-832.
38. Fitch, A.; Wang, Y.; Mellican, S.; Macha, S. *Anal. Chem.* **1996**, *68*, 727A-731A.
39. Wilson, G. S.; Anderson, M. R.; Lunte, C. E. *Anal. Chem.* **1999**, *71*, 677A-681A.
40. Hope, W. W.; Johnson, L. P. *Anal. Chem.* **2000**, *72*, 460A-467A.
41. Thompson, R. Q.; Edmiston, P. L. *Anal. Chem.* **2001**, *73*, 678A-684A.
42. Walters, J. P. *Anal. Chem.* **1991**, *63*, 977A-985A; *Anal. Chem.* **1991**, *63*, 1077A-1087A; *Anal. Chem.* **1991**, *63*, 1179A-1191A.
43. Juhl, L.; Yearsley, K.; Silva, A. J. *J. Chem. Educ.* **1997**, *74*, 1431-1433.
44. Ross, M. R.; Fulton, R. B. *J. Chem. Educ.* **1994**, *71*, 141-143.
45. Peters, A. W. *J. Chem. Educ.* **2005**, *82*, 571-574.
46. Craig, P. A. *J. Chem. Educ.* **1999**, *76*, 1130-1135.
47. Dunn, J. G.; Phillips, D. N. *J. Chem. Educ.* **1998**, *75*, 866-869.
48. Bradley, A. Z.; Ulrich, S. M.; Jones, Jr., M..; Jones, S. M. *J. Chem. Educ.* **2002**, *79*, 514-519.
49. Browne, L. M.; Blackburn, E. V. *J. Chem. Educ.* **1999**, *76*, 1104-1107.
50. Davis, D. S.; Hargrove, R. J.; Hugdahl, J. D. *J. Chem. Educ.* **1999**, *76*, 1127-1130.
51. Paulson, D. R. *J. Chem. Educ.* **1999**, *76*, 1136-1140.
52. Hagen, J. P. *J. Chem. Educ.* **2000**, *77*, 1441-1444.
53. Hass, M. A. *J. Chem. Educ.* **2000**, *77*, 1035-1038.
54. Deckert, A. A.; Nestor, L. P.; DiLullo, D. *J. Chem. Educ.* **1998**, *75*, 860-863.
55. Towns, M. H.; Kreke, K.; Fields, A. *J. Chem. Educ.* **2000**, *77*, 111-115.
56. Hinde, R. J.; Kovac, J. *J. Chem. Educ.* **2001**, *78*, 93-99.
57. Wenzel, T. J. *CBE – Life Sci. Educ.* **2006**, *5*, 76-84.
58. Wenzel, T. J. *Gender, Science and the Undergraduate Curriculum: Building Two Way Streets* **2001**, 29-46, Association of American Colleges and Universities, Washington, DC.
59. Wenzel, T. J. *J. Chem. Educ.*, **2007**, *84*, 182-186.
60. The cooperative learning exercises are available at http://www.bates.edu/x65385 and http://www.bates.edu/x65478 (accessed on October 27, 2006).
61. Wenzel, T. J. *Anal. Chem.* **2002**, *74*, 439A-440A.
62. Thoreau, H. D. *Walden: Or, Life in the Woods* **1995**, Dover, Mineola, NY.

Chapter 6

Bioanalytical Chemistry: Model for a Fully Integrated Problem-Based Learning Approach

Patricia Ann Mabrouk

Department of Chemistry and Chemical Biology, Northeastern University, 360 Huntington Avenue, Boston, MA 02115

This chapter describes an advanced analytical undergraduate elective at Northeastern University entitled "Bioanalytical Chemistry," which is based on the problem-based learning (PBL) model of instruction. Bioanalytical Chemistry has been offered three times (2001, 2004, and 2006) since 2001 to classes ranging in size from four to fourteen undergraduates. In this semi-structured experiential learning course, students work for a semester in teams of 3-5 members on genuine analytical research problems using research-grade analytical instrumentation. A number of learning techniques including weekly group meetings, job safety analyses, and reflective journals provide supportive scaffolding for students and the instructor, whose principal role is that of a facilitator and coach. Summative assessment data for the period 2001-2006 document the efficacy of this approach in promoting higher level cognitive learning and provide insight into how and why this methodology works.

Problem-based learning (PBL) is an instructional method that emphasizes cooperative work in small groups on real world problems as a mechanism to engage student interest and promote deeper student learning. PBL was originally introduced and popularized in the early 1980's by the medical community as a method of instructing future physicians (*1*). Subsequently the method began to be adapted and adopted by other academic disciplines in the mid-1990's including chemistry (*2-6*). The approach found strong support within the analytical chemical education community (*2,7-9*). PBL is usually practiced either in the classroom or in the laboratory. Several chapters in this book discuss classroom or laboratory experiments based on this model. The present chapter will focus on an integrated classroom-laboratory model in which efforts in both the classroom and the laboratory center on the investigation of a genuine bio-analytical research project.

By way of background, I first introduced PBL into the chemistry laboratory curriculum at Northeastern University for the Quantitative Analysis course in 1994 when asked to update the laboratory curriculum. At that time I created a series of multi-week bio-analytical laboratory projects introducing undergraduates to several modes of modern analytical separations. These experiments replaced a number of the titrimetry experiments using Thorn-Smith samples that dominated the laboratory curriculum for this course. Several of these PBL-based projects were subsequently described in the peer-reviewed literature (*10-12*). Five years ago in response to student interest in an advanced undergraduate analytical elective using the PBL methodology and the bio-sciences and a departmental interest in an advanced science elective for biochemistry and engineering majors, Bioanalytical Chemistry (CHM U331) was introduced into the undergraduate chemistry curriculum (*13*).

The new semester-long course has been offered three times (spring semester 2001, 2004, and 2006) since it was added to the curriculum in 2001 and is now offered once annually. Bioanalytical Chemistry counts as an advanced science elective for science and engineering majors. The only pre-requisite is Quantitative Analysis, which is typically taken by biochemistry and biology majors immediately following the standard two-semester introductory chemistry sequence. To date undergraduates electing this course have been primarily chemistry, biochemistry, and chemical engineering majors. Course enrollment has grown from four to fourteen students. The course could likely accommodate more students however the availability of analytical instrumentation will ultimately limit the course size.

In Bioanalytical Chemistry students, working in self-directing teams of three to five learners, investigate an authentic analytical research problem that they identify. Examples of past projects include investigation of the use of differential scanning calorimetry as a method of evaluating the quality of chocolate, the isolation and purification of soybean peroxidase from soybean hulls, the gas chromatography-mass spectrometry (GC-MS) analysis of neutraceuticals such as gamma amino butyric acid or the adulteration of henna with *p*-phenylene diamine (black henna) by HPLC. The focus throughout the course is on process of chemical analysis rather than the acquisition of specific technical content.

Course Format

As mentioned earlier, PBL is usually employed either in the classroom or in the laboratory on one or more discrete projects or experiments. When applied in this way, there is no need to modify the traditional course format. However, in the current application since accomplishment of the students' experimental research projects drives the class, the traditional course format, which at Northeastern University consists of a series of three 65-min class meetings and one 3-h laboratory meeting each week, may, at first, appear less than ideal. However, it is no easier at Northeastern University than it is at any other college or university to offer courses with a non-standard format. Adoption of a non-traditional course format, while possible, proved impractical due to the unpredictability, inflexibility, and complexity of individual student course schedules. Consequently, I have attempted to make the best use of the traditional science course format. Two of the three lecture meetings serve as classroom discussions of student-identified topics related to perceived needs for information on the group research projects. The third classroom meeting functions as a weekly "group meeting" for the student teams. Group meetings are modeled after graduate research group meetings; each team presents a brief PowerPoint®-based progress report summarizing their accomplishments in the laboratory and outlining their laboratory plans for the upcoming week. Because laboratory time is limited, teams quickly learn to maximize their productivity by dividing up the work load and multi-tasking, *i.e.*, positive interdependence between team members is fostered and students learn invaluable time management and project management skills.

Student Project Selection

On the first day of the course, students are provided with a list of several past research projects (models) as well as available research-grade analytical instrumentation. The undergraduates are asked to form teams and to identify a research project, of interest to the group, using the available analytical instrumentation, which they feel they can accomplish over a 10-week period.

Over the next week or two, the students work together to define their research project, read the relevant technical literature, identify a hypothesis which they will test, specify the analytical techniques that they will use in order to investigate the problem, and to begin to identify technical content needs. To ensure that the students don't waste the limited laboratory time pursuing dead ends, teams present weekly oral reports beginning the first week of the course. This allows me to provide timely intervention, if required, and the environment establishes a healthy competitive spirit and open, free flowing communication within and between teams that promotes and enriches student learning.

Early on I learned that the projects that work best are those that the students themselves craft. This promotes student ownership not only of the project but of the learning experience as a whole. These projects need not be "perfect" in their design. In fact, the students appear to learn more when the project is imperfect (*14*). For example, last year a group of students originally expressed an interest in using GC-MS to quantitate the amount of pesticides in ground beef. After encouraging the group to think about the amount and types of pesticide that might be present, sampling issues, and detection limits for the mass spectrometer, the group decided instead to quantitate the omega-6 fatty acids in beef using GC-MS. Their goal was determination of the relative amount of the omega fatty acids in ground beef from organic/grass fed steers as versus grain-fed steers with the expectation that grain-fed beef would contain higher levels of omega-6 fatty acids (*15*). In the end, the students learned a great deal about the biochemistry of lipids, organic chemistry (they prepared and analyzed methyl ester derivatives) (*16*) as well as analytical chemistry (GC and mass spectral analysis). What is important in terms of project selection and design is that the students identify an authentic problem, one whose solution the students believe is likely to be of current interest to the greater scientific community (student knowledge of the relevant peer-reviewed literature and the relevant scientific community is admittedly limited), suitably sophisticated in its demands on the students' cognitive and laboratory skills, and one for which the needed supports (instrumentation, materials and supplies, and instructor assistance) can be supplied. In this way, the student learners own not only the selection of the project but equally importantly the problem solving process and the eventual solution of the research problem.

When I began teaching this course I quickly realized that it could be quite expensive for the department and excessively demanding and time consuming for me if I didn't guide my students' project selection somewhat based on the available instrumentation, the cost of the required reagents and supplies such as GC or HPLC columns, etc. Consequently, students are asked to informally cost out the materials and supplies required for their project at the start and to include this information in their first formal group meeting presentation. An alternative strategy might be to require student teams to submit a formal proposal and budget (*9*). As the course already includes several significant writing assignments (reflective journals, job safety analysis (JSA), a standard operating protocol (SOP) and a final technical paper), more than the norm in a science or engineering course, I have not used a formal proposal as a scaffold support.

The last critical element in project selection relates to the instructor him/herself. To make effective use of this approach, the instructor must be prepared to give students ownership not only of the project selection but also of the project design – the problem solving process. The instructor must serve as consultant or coach and be careful not to assume control of the problem solving process (*17*). This means that the instructor must recognize his/her own areas of self-avowed technical expertise and his/her own tolerance of the unknown. Finally, but perhaps most importantly, the instructor must value his/her students' thinking, *i.e.*, the instructor must be savvy in determining the appropriate degree of sophistication that his/her students can handle in terms of project difficulty.

Groups and Group Structure

Collaborative inquiry by teams of students is a distinctive feature of the problem-based learning model. The benefits of small group learning are well established (*18,19*). Positively structured, small groups produce greater student learning, achievement, satisfaction and increased persistence through required academic coursework (*17,19*).

The creation of productive student teams is critical in facilitating productive learning using this model. Consequently a number of different approaches have been taken to facilitate effective teamwork through the deliberate identification of groups, the assignment of distinct group roles, and the imposition of ground rules for group function (*5*). Creation of heterogeneous groups of students (with regard to gender, ethnicity, and academic performance) by the instructor is frequently advocated in the literature (*5,20*). This ensures that teams are balanced in terms of strengths, weaknesses, experience, and perspectives and the mix maximizes everyone's learning opportunities (*21*). Role play can be a useful mechanism to facilitate and support group work on complex problems (*22*). One such approach was popularized by John Walters, Professor Emeritus at St. Olaf College, who organized students in his instrumental analysis course into groups of four students called "companies" in which each student had a prescribed, tightly defined role such as manager, hardware, and chemist (*23-25*). Students are rotated through the different roles on the various laboratory experiments run over the semester.

In Bioanalytical Chemistry, I encourage students to identify their own teams based on their interests the first week of the course. I do not define group size but encourage students to work in groups of 3-5 students. Like others (*20*), I have found that teams of two or four (even numbers) students seem to work best. Pairs of students can identify distinct roles and tasks. Similarly, groups of four students can break into pairs of students who can partner to carry out unique roles and tasks in the laboratory. If you use larger teams, however, some students may not find their voice within the group and equitable distribution of tasks may become an issue.

However, you choose to do it, it is vital that students form productive self-directed groups. This means the students must develop positive working relationships with unique and clearly defined roles and responsibilities, and rules for working together as a unit. I prefer a less structured and more student-centered and student-directed approach to the development of productive, collegial teams that encourages students to define their roles, work policies, etc. within their groups. I see this as an important element/aspect of successful team development. A useful artifice supporting effective self-directed team formation are group covenants (*26*) or group contracts (*27*). Group covenants, negotiated within the first two weeks of the semester, require teams to identify individual roles and responsibilities and to discuss and deal with individual members' beliefs regarding the advantages and disadvantages of working together as a team. If issues arise concerning individual member workload, responsibilities, etc. group covenants help teams adjudicate these issues themselves again encouraging self-direction.

Scaffolded Supports for Learning

I use a variety of mechanisms to support student learners' inquiry and to provide timely individual and group feedback concerning performance. Some are standard approaches familiar to chemistry students and instructors alike such as a midterm examination, group oral presentations, SOPs, peer evaluation, and submission of an independently authored technical paper (*Anal. Chem.* format). However, a number are new to my students and likely unfamiliar to many chemistry faculty as well in the way that I execute them. These include discussions, structured reflective journaling, and JSAs. In the following sections we'll review these mechanisms in terms of the function that each performs in promoting student learning and facilitating instructor assessment of the quality of individual and group student learning. It is important to note that there are many other useful mechanisms you can use and that you can adapt these to suit your specific needs. Wilson *et al.* (*28*) conduct monthly oral and written job performance evaluations with each student that includes both individual and group performance elements and hold a final "International Symposium on Instrumental Analysis Laboratory." Kalivas uses full research proposals with a budget and SOPs in support of his PBL efforts (*9,29*).

Classroom Discussions

The classroom environment in Bioanalytical Chemistry is genuinely constructivistic. The course does not have an instructor-predefined and driven course curriculum. In Bioanalytical Chemistry, student project needs drive the selection, timing, and content of *discussions* in the course. Discussion topics are identified by the class in the discussion sessions. Topics may be requested by the class with a minimum lead time of 1-day advance notice to the instructor. These sessions are not traditional lectures but rather are discussions; their purpose is not to provide the principal/primary explication on a topic but rather to serve as active, student-centered problem solving sessions. They provide a forum for focused, group discussion of topics of confusion and to ensure that students are properly resourced for authentic inquiry.

This approach facilitates student learning as the students have identified their content needs, have made an effort to understand and process the information, and have often identified specific stumbling blocks with which they need external assistance. This allows the instructor the ability to provide timely, targeted feedback and assistance without wrestling control of the problem solving and critical thinking away from the students.

This method might sound chaotic and haphazard – perhaps even fool hardy from a course curriculum viewpoint. What is most intriguing is that, though not predefined, there really is a core curriculum driving Bioanalytical Chemistry. The specifics do change depending on the backgrounds of the students and the specific projects selected. However, over the past five years I have observed a surprising degree of consistency in terms of the general topics and the order in which these topics are covered. Discussion topics usually include: statistics, sampling, quality control and assurance standards, troubleshooting strategies, effective bibliographic search strategies, the format of the technical journal article, the basics of relevant instrumentation, etc. The order and timing of topics reflects student needs in terms of their research projects. Student researchers often wrestle at the beginning with sampling. In part this reflects the limited treatment of sampling in the Quantitative Analysis course (*30*). Laboratory experiments for the introductory analytical course usually emphasize analysis of analytes in simple matrices like water, and textbooks don't discuss sampling issues in any significant detail. Subsequently, student focus usually shifts to the analytical instrumentation they plan to use. They are typically focused on learning to use the instrumentation and subsequently their focus shifts to an interest in the underlying theory. As students begin to acquire experimental data on their projects they become interested in the quality of their data, specifically, how to demonstrate the quality of their work to others. This usually culminates in an interest in understanding how to effectively communicate their findings to others in the field of analytical chemistry. In short, the course curriculum mirrors the stages in the modern bio-analytical analysis.

Reflective Journaling

An important objective of the course is the development of self-directed student analysts. A powerful tool widely used in the social sciences and in the health professions but not commonly used in chemistry instruction at colleges or universities facilitating self-direction is the practice of reflective journaling (*31-33*). The practice of structured reflective journaling is a form of reflective journaling that is guided by a series of questions intended to facilitate reflection on an experience or series of experiences. With judicious selection of guiding questions, the practice has been shown to promote self-directed understanding of one's analytical thinking, problem solving, and critical reflection skills while at the same time enhancing one's connectedness and sense of community (*34,35*). Journals accomplish this by helping the student to make his/her thinking explicit and therefore apparent to themselves. This enables the student to better understand his/her assumptions, biases, thought processes, etc. and therefore to identify and self-evaluate possible solutions to the problems they identify in their journals which in turn encourages long term self-directed personal growth. The imposition of structure onto the journaling exercise in the form of guiding questions can be particularly beneficial to novice practitioners who may lack the skills and discipline to analyze their workplace experiences.

In "Bioanalytical Chemistry," I ask the students to journal throughout the semester and to submit to the instructor and their peers three brief (1-page), typewritten reflective journal entries over the course of the semester. Journal entries are purposely distributed to classmates as public reflection has been argued to deepen individual and group learning by encouraging open, transformative discourse (*36,37*). I do not use a prescriptive format for these assignments but rather offer my students a series of six questions to guide students in their reflection. Identification and selection of these questions has been guided by the literature on reflective journaling and critical reflection (*37*). The questions I have chosen ask students to identify an experience (content reflection), own their assumptions, prejudices and/or biases (premise), place the experience in perspective, and to generate and to constructively evaluate possible solutions to the problem with an eye toward the future (process).

Concurrent structured reflective journaling has proven to be an extremely useful tool in teaching my students the value of good interpersonal (listening, sensitivity, conflict resolution, etc.), critical thinking and problem solving skills. More importantly, I believe my students gain a deeper understanding of themselves (personal values, world view, strengths and weaknesses), increased self-direction, a stronger professional identity and a healthy self-confidence in their technical abilities. My perceptions are based on the quality and depth of reflection I see in their journal entries.

Job Safety Analyses (JSAs)

Job safety analyses are frequently used by government laboratories and private industry (*38*). JSAs outline the potential dangers, if any, represented by each experimental operation that an investigator plans to carry out in the laboratory when executing a particular experimental protocol together with the safeguards in place to prevent accidents. By their very nature JSAs can be very useful to an instructor supervising team-centered undergraduate research projects. JSAs encourage thoughtful, step-by-step planning. Of course, advance planning helps the students make efficient use of the 3-h laboratory available to them each week. To ensure individual as well as group accountability, all team members must sign the JSA before submitting it to me. Finally, JSAs must be submitted at least one day prior to the lab. This ensures that I have adequate time to review and discuss the group's plans with them, if needed, in advance of the laboratory session.

Reward Structure

Another important element of course design in using cooperative learning methods centers on the reward structure (*39*). Grades should value both individual and group performance (*20*). Grading (*40*) serves three purposes in Bioanalytical Chemistry: communication, motivation, and evaluation. Criteria for evaluation have been identified to provide students with timely feedback concerning both individual and group performance. Specific evaluative measures include some traditional elements as well as some non-traditional elements: a written midterm examination (15% of final course grade), group meeting oral presentations (average of six; 10%), job safety analysis reports (weekly; 10%), group SOP (10 %), group notebook checks (average of two; 10%), individual reflective journals (average of three; 10%), team contribution grade (5%), final group presentation (20%), and final technical report (20%). Whenever possible, students are provided in advance with rubrics (*41,42*) outlining the standards by which their performance will be evaluated. This ensures that assignments are worth grading both from the standpoint of the student learner as well as that of the instructor.

Summative Course Evaluation

Two survey tools have been consistently employed to evaluate the Bioanalytical chemistry course. First, a traditional institutionally designed student course evaluation survey referred to here as "TCEPs." In addition, a modified version of Elaine Seymour's Student Assessment of Learning Gains (SALG) tool (URL: http://www.wcer.wisc.edu/salgains/instructor/) was used. This anonymous, on-line opinion survey tool provides insight into student perceptions of their learning gains. Application for human subjects research was filed with the Northeastern University Institutional Review Board in 2001 for use of the SALG survey tool and approved as DHHS exempt, category 1. Both survey tools use Likert scales and allow students to write brief narrative responses as well.

Except in one instance, surveys were administered approximately one week prior to the end of the course and were strictly anonymous. In 2006, the SALG was administered approximately two months after the course had ended. In all cases, return rates were above 50%. Response rates were notably higher for the TCEPs than for the SALG surveys. The TCEPs were administered in class near the end of a class period while the SALG surveys, which had to be completed on-line, were completed outside of class at a time of the student's choosing. Differences in the return rates may therefore reflect the extra effort required to complete the SALG survey outside of class time coupled with a possible perception of the redundancy of the SALG survey from a student perspective. Selected TCEP and SALG survey data for 2001-2006 are summarized in Tables 1-5.

TCEP Data

Overall, as reflected in the TCEP data shown in Table 1, students consistently rank the Bioanalytical Chemistry course quite highly both in terms of the overall course rating as well and the amount of information learned.

Student narrative comments from 2001 to 2006 support the students' strong approval of this course:

"A great class (one of the best i've taken)..."

"The class was a great experience..."

"Everything was approached differently than all the lab courses ive ever had. It was wonderful."

Student satisfaction is particularly noteworthy as (see Table 1) the students perceive Bioanalytical Chemistry to be a challenging course.

Table 1. Summary of NU TCEP Student Ratings Data Reports for Bioanalytical Chemistry

Question	*2001 (4/4)*[a]		*2004 (14/12)*		*2006 (8/8)*	
	Mean[b]	*SD*[c]	*Mean*	*SD*	*Mean*	*SD*
Difficulty level of the course[d]	3.3	0.5	3.3	0.7	4.0	0.5
Usefulness of outside assignments	4.0	0.0	4.2	0.8	4.1	0.6
Usefulness of in-class activities	4.5	0.6	4.4	0.8	4.5	0.8
How much have you learned in this course	4.5	0.6	4.0	1.0	4.3	0.5
Overall rating of this course	4.3	1.0	4.4	0.8	4.3	0.5

[a]Number of students enrolled/Number of students evaluated

[b]Ratings based on 5-point Likert scale: 5 (strongly agree), 4 (agree), 3 (neutral), 2 (disagree), and 1 (strongly disagree)

[c]Standard deviation

[d]Ratings based on 5-point Likert scale: 5 (among the most difficult), 3 (about average), and 1 (among easiest)

Course-related activities both inside and outside the classroom are consistently rated very highly by students (see Table 1). Since the students themselves identify the topics for in class discussion and weekly group meetings, perhaps it is not surprising that they rate the classroom activities quite favorably. However, narrative comments suggest that student satisfaction with classroom activities may be in part due to the integrity of the learning experience, achieved through the integration of the laboratory and classroom:

"Everything was approached differently than all the lab courses ive[sp.] ever had. It was wonderful. Lab was an integrative part of the learing[sp.] experience"

"I also feel that learning from doing always carries with you longer than listening in class."

"class backed up lab work"

Students spend a significant amount of time working individually and with their teammates on a number of assignments and activities including the reading the relevant primary technical literature, planning experiments, analyzing data, and preparing JSAs, etc. which must be done outside the classroom. Many of these assignments have been designed to serve as discussed earlier as scaffolding supports. Thus, it is notable that student respondents consistently rate the outside assignments highly as well (Table 1). Narrative comments offered on both the TCEP and SALG surveys suggest that the students are willing to put in the extra effort because they feel it is worthwhile. For example, one student wrote: "This class is definitely worth the time and effort. There is much to gain in independent work with moderate guidance."

SALG Data

The SALG survey was used to gain insight into what (1) the students felt they learned (knowledge, understanding, skills, etc.) (see Tables 2 and 3); (2) to ascertain to what extent students felt they made learning gains (see Table 4); and to (3) determine whether or not the students felt there was long term value in what they had learned in the course (see Table 5). The relevant data are summarized in Tables 2-5.

The student respondents in Bioanalytical Chemistry consistently indicated strong gains in their understanding in all of the areas queried on the SALG survey (see Tables 2 and 3), specifically, in learning how to use new/different analytical instrumentation, in experimental design, troubleshooting, data analysis, reading the primary technical literature, making scientific presentations, writing technical papers, and teamwork. The respondents also indicated gains in their experimental design, oral communications, critical thinking, and teamwork skills. However, the rating data for teamwork are the lowest and exhibit the greatest variations likely reflecting the degree of satisfaction of students with their experience with their particular research group.

Students felt that they made positive gains in all of the areas queried including knowledge, critical thinking, and self-confidence (see Table 4). However, the greatest gains were perceived to be in the students' appreciation of

Table 2. Summary of SALG Student Perception Rating Data Reports for Bioanalytical Chemistry

Q: As a result of your work in this class, how well do you think that you now understand each of the following?

Question	*2001 (4/3)*[a]		*2004 (14/8)*		*2006 (8/4)*	
	Mean[b]	*SD*[c]	*Mean*	*SD*	*Mean*	*SD*
How to use a new/different analytical instrument?	4.0	0.0	4.4	0.7	4.0	1.0
How to design an experiment in order to obtain meaningful information?	4.5	0.5	4.4	0.7	4.0	1.0
How to troubleshoot a problem with an instrument or method?	4.0	0.0	4.4	0.7	3.3	0.4
How to interpret data?	3.5	0.5	4.3	0.4	3.8	0.4
How to read and understand the primary literature?	4.0	0.0	4.6	0.5	4.0	1.2
How to make an effective scientific presentation?	4.0	0.0	4.6	0.5	4.0	1.2
How to work with others in a team?	4.0	0.0	4.8	0.4	3.8	1.1
How to write a technical paper?	4.0	0.0	4.5	0.7	3.8	1.1

[a]Number of students enrolled/Number of students evaluated

[b]Ratings based on 5-point Likert scale: 5 (strongly agree), 4 (agree), 3 (neutral), 2 (disagree), and 1 (strongly disagree)

[c]Standard deviation

Table 3. Summary of SALG Student Perception Rating Data Reports for Bioanalytical Chemistry

Q: How much has this class added to your skills in each of the following?

Question	*2001 (4/3)*[a]		*2004 (14/8)*		*2006 (8/4)*	
	Mean[b]	*SD*[c]	*Mean*	*SD*	*Mean*	*SD*
Troubleshooting?	4.0	0.0	4.38	0.7	3.25	0.43
Writing papers?	4.5	0.5	4.0	0.87	3.75	0.43
Designing lab experiments?	4.0	0.0	4.5	0.71	4.0	1.22
Finding trends in data?	4.0	0.0	3.75	0.97	4.25	1.3
Working effectively with others?	5.0	0.0	4.25	0.83	3.5	1.12
Giving oral presentations?	4.0	0.0	4.38	0.86	4.0	1.22
Critical thinking?	5.0	0.0	4.38	0.7	4.0	1.22

[a]Number of students enrolled/Number of students evaluated

[b]Ratings based on 5-point Likert scale: 5 (a great deal), 4 (a lot), 3 (somewhat), 2 (a little), and 1 (nothing)

[c]Standard deviation

the field of analytical chemistry and in their understanding of its relevance to the real world. Both skills represent examples of the evaluation competency, the most complex of the six skill categories, in Bloom's taxonomy of cognitive learning (*43*). This suggests that the PBL learning process has been effective in promoting higher level cognitive skill development.

Finally, students perceive there to be significant long term value in what they feel that they have learned in the course (see Table 5). Perhaps not surprisingly, students appear to feel most strongly that they carry away what they learned working on their project. This is a clear statement of the unique transformative power of experiential learning. Of course the key question is how this is achieved. A number of narrative comments provide insight into a possible mechanism, specifically, role play. Participation in the solution of a genuine research problem provides students the opportunity to test and authenticate their professional self-identity:

> "Overall, I thought that this was a great course, and I learned a great deal about myself personally and professionally. I discovered some hidden skills, as well as some weaknesses that I got to work on."
>
> "I acheived[sp.] much greater confidence (in terms of discussing and speaking about) in talking about my major and its related fields"

Table 4. Summary of SALG Student Perception Rating Data Reports for Bioanalytical Chemistry

Q: To what extent did you make gains in any of the following as a result of what you did in this class?

Question	*2001 (4/3)*[a]		*2004 (14/8)*		*2006 (8/4)*	
	Mean[b]	*SD*[c]	*Mean*	*SD*	*Mean*	*SD*
Understanding the main concepts	4.0	0.0	4.25	0.83	3.5	0.87
Understanding the relationship between concepts	4.0	0.0	4.25	0.66	3.5	1.12
Understanding how ideas in this class relate to those in other science classes	3.5	0.5	4.71	0.7	3.5	0.5
Understanding the relevance of this field to real world issues	4.0	0.0	4.62	0.48	4.75	0.43
Appreciating this field	4.0	0.0	4.12	0.93	4.25	0.43
Ability to think through a problem or argument	4.0	0.0	4.38	0.7	3.5	0.87
Confidence in your ability to do work in this field	3.5	0.5	3.75	0.97	3.5	1.5
Feeling comfortable with complex ideas	4.0	0.0	4.29	0.7	3.5	1.5
Enthusiasm for subject	4.0	0.0	3.62	0.86	3.5	1.5

[a]Number of students enrolled/Number of students evaluated

[b]Ratings based on 5-point Likert scale: 5 (a great deal), 4 (a lot), 3 (somewhat), 2 (a little), and 1 (not at all)

[c]Standard deviation

"This was a great upper level class to really put the feild[sp.] into real world and professional perspective for me. I enjoyed the class and it made me feel more like a "scientist" :)"

Table 5. Summary of SALG Student Perception Rating Data Reports for Bioanalytical Chemistry

Q: How much of the following do you think you will remember and carry with you into other classes or aspects of your life?

Question	*2001 (4/3)*[a]		*2004 (14/8)*		*2006 (8/4)*	
	Mean[b]	*SD*[c]	*Mean*	*SD*	*Mean*	*SD*
What I learned in lecture	3.0	0.0	4.0	1.0	3.3	0.8
What I learned working on my project	4.5	0.5	4.4	0.7	4.5	0.5
Personal relationships I developed with other students in the course	3.0	0.0	4.1	0.8	3.0	1.4
Confidence in my ability to learn new and difficult things	4.0	0.0	4.4	0.9	3.5	0.9
Confidence in my ability to communicate effectively	4.0	0.0	4.5	0.7	3.5	1.5
Problem-solving skills	4.0	0.0	4.3	0.8	3.3	0.8
Lab technique	3.5	0.5	4.5	0.7	3.5	0.5
Time management skills	4.0	0.0	4.5	0.7	4.0	0.7
Troubleshooting skills	4.0	0.0	4.3	0.7	3.8	0.4
Technical writing[d]	---	---	4.5	0.7	4.0	0.7

[a]Number of students enrolled/Number of students evaluated

[b]Ratings based on 5-point Likert scale: 5 (a great deal), 4 (a lot), 3 (somewhat), 2 (a little), and 1 (not at all)

[c]Standard deviation

[d]Students were not queried on this aspect of the course in 2001

Conclusions

A fully integrated problem-based learning model suitable for adaptation and adoption by other science, technology, engineering, and mathematics disciplines has been articulated. Preliminary assessment data suggest the model is an extremely effective method of teaching and learning promoting more complex cognitive development within the learner. Didactic knowledge becomes contextualized and enriched through the intellectual exchange of ideas by interested, fully-engaged team members Knowledge is tested, analyzed, synthesized and evaluated, *i.e.*, students make use of the full range of cognitive competencies. Perhaps most importantly, the approach appears to provide an environment to test and affirm their professional self-identity.

References

1. Barrows, H. S.; Tamblyn, R. *Problem-Based Learning: An Approach to Medical Education*; Springer: New York, 1980.
2. Wenzel, T. J. *Anal. Chem.* **1995**, *67*, 470A-475A.
3. *Bringing Problem-Based Learning to Higher Education: Theory and Practice*; Wilkerson, L.; Gijselaers, W. H., Eds.; Jossey-Bass: San Francisco, 1996; Vol. 68.
4. Duch, B. *J. Coll. Sci. Teach.* **1996**, 326-329.
5. *The Power of Problem-Based Learning*; Duch, B. J.; Groh, S. E.; Allen, D. E., Eds.; Stylus: Sterling, 2001.
6. Allen, D.; Tanner, K. *Cell Biol. Educ.* **2003**, *2*, 73-81.
7. Kuwana, T. "Curricular Developments in the Analytical Sciences," National Science Foundation, 1998.
8. Ram, P. *J. Chem. Educ.* **1999**, *76*, 1122-1126.
9. Kalivas, J. H. *J. Chem. Educ.* **2005**, *82*, 895-897.
10. Mabrouk, P. A. *J. Chem. Educ.* **1996**, *73*, A149-A152.
11. Marzilli, L. A.; Bedard, P.; Mabrouk, P. A. *Chem. Educ.* **1997**, *1*, 1-12.
12. Mabrouk, P. A.; Castriotta, K. *J. Chem. Educ.* **2001**, *78*, 1385-1386.
13. Mabrouk, P. A. *Anal. Bioanal. Chem.* **2005**, *383*, 365-367.
14. Wenzel, T. J. *Anal. Chem.* **2002**, 439A-440A.
15. Enser, M.; Hallett, K. G.; Hewett, B.; Fursey, G. A. J.; Wood, J. D.; Harrington, G. *Meat Sci.* **1998**, *49*, 329-341.
16. Long, G. L. *J. Chem. Educ.* **1975**, *52*, 813-814.
17. Cohen, E. G. *Rev. Educ. Res.* **1994**, *1994*, 1-35.
18. Qin, Z.; Johnson, D. W.; Johnson, R. T. *Rev. Educ. Res.* **1995**, *65*, 129-143.
19. Springer, L.; Stanne, M. E.; Donovan, S. S. *Rev. Educ. Res.* **1999**, *69*, 21-51.
20. Cooper, M. M. *J. Chem. Educ.* **1995**, *72*, 162-164.
21. Michaelson, L. K.; Black, R. H. In *Collaborative Learning: A Sourcebook for Higher Education*; Kadel, S., Keener, J., Eds.; National Center for Teaching and Learning Assessment: State College, 1994; Vol. 2.
22. Deavor, J. P. *J. Chem. Educ.* **1994**, *71*, 980-982.
23. Walters, J. P. *Anal. Chem.* **1991**, *63*, 977A-985A.
24. Walters, J. P. *Anal. Chem.* **1991**, *63*, 1077A-1087A.
25. Walters, J. P. *Anal. Chem.* **1991**, *67*, 1179A-1191A.

26. Towns, M. H. *J. Chem. Educ.* **1998**, *75*, 67-69.
27. Crowe, M.; Hill, C. *J. Coll. Sci. Teach.* **2006**, *35*, 32-35.
28. Wilson, G. S.; Anderson, M. R.; Lunte, C. E. *Anal. Chem.* **1999**, *71*, 677A-681A.
29. Houghton, T. P.; Kalivas, J. H. *J. Chem. Educ.* **2000**, *77*, 1314-1318.
30. Sherren, A. T. *J. Chem. Educ.* **1991**, *68*, 598-599.
31. *Promoting Journal Writing in Adult Education*; English, L. M.; Gillen, M. A., Eds.; Jossey-Bass: San Francisco, 2001; Vol. 90.
32. Etkina, E.; Harper, K. A. *J. Coll. Sci. Teach.* **2002**, *31*, 476-480.
33. Peery, A. B. *ARRIVE: Improving Instruction through Reflective Journaling*; Advanced Learning Press: Englewood, 2005.
34. Kahaney, P.; Heinrich, K. In *Interdisciplinary Handbook of Adult Lifespan Learning*; Sinnott, J. D., Ed.; Greenwood Press: Westport, 1994.
35. Blake, T. K. *Int. J. Nurs. Educ. Scholarsh.* **2005**, *2*, article 7.
36. Mezirow, J. *Fostering Critical Reflection in Adulthood: A Guide to Transformative and Emancipatory Learning*; Jossey-Bass: San Francisco, 1990.
37. Raelin, J. A. *Management Learning* **2001**, *32*, 11-30.
38. Seidman, L. A.; Moore, C. J. *Basic Laboratory Methods for Biotechnology: Textbook and Laboratory Reference*; Prentice-Hall: Upper Saddle River, 2000.
39. Webb, N. M. *Rev. Educ. Res.* **1982**, *52*, 421-445.
40. Walvoord, B. E.; Anderson, V. J. *Effective Grading*; Jossey-Bass: San Francisco, 1998.
41. Lunsford, E.; Melear, C. T. *J. Coll. Sci. Teach.* **2004**, *34*, 34-38.
42. Walvoord, B. E. *Assessment Clear and Simple*; Jossey-Bass: San Francisco, 2004.
43. Bloom, B. S.; Krathwohl, D. R. *Taxonomy of Educational Objectives, Handbook I: Cognitive Domain*; Longmans: New York, 1956.

Chapter 7

Context-Based Case Studies in Analytical Chemistry

S. T. Belt[1] and T. L. Overton[2]

[1]Centre for Chemical Sciences, School of Earth, Ocean and Environmental Sciences, University of Plymouth, Drake Circus, Plymouth PL4 8AA, United Kingdom
[2]Department of Chemistry, University of Hull, Hull HU6 7RX, United Kingdom

This paper describes an overview of the development and use of a series of context-based case studies for the teaching of analytical chemistry to undergraduate students. A rationale behind using case studies in teaching analytical chemistry is given, together with a description of the case studies that we have developed and use with our students. Two of these are described in more detail. We summarize our experiences in using a context-based approach to teaching analytical chemistry including feedback from students.

Introduction

In the United Kingdom, educators and funding bodies continue to highlight the importance of the development of a wide range of subject and transferable skills during students' higher education careers (*1-3*). These views have been endorsed, in particular, within the analytical sciences (*4*). In addition, employers have identified numerous key skills that are important, but are often not well developed in recent graduates (*5*). As a result, funding bodies have identified specific subject and transferable skills as essential components that students must achieve as part of their courses. In order to assist educators in designing courses, the Quality Assurance Agency (QAA) has produced guidelines through the

Subject Benchmarking activity and the Programme Specification Template where the full range of desirable skills are identified (*6,7*). In addition, and as part of their quality assurance procedures, university chemistry departments will need to demonstrate that their courses enable students to develop such skills and be assessed in them (i.e. they are integral components of the courses). Of course, defining an exhaustive list of subject and transferable skills relevant to a modern chemistry graduate is a near impossible task though some broad classifications are probably helpful. For example, a graduate in chemistry needs to be able to collect and interpret chemical information and present their findings in an appropriate scientific format. In doing this, they will frequently encounter novel or unfamiliar scenarios which require them to problem-solve and plan strategies for their solution. From a logistical standpoint, students will need to work with others, engage in various information retrieval activities and manage their time effectively. Our view has been that while no single mode of teaching is capable of satisfying all of the learning skills required of chemistry students, the use of context-based case studies encourages various skills types to be developed in a coherent and integrated manner. Further, the use of case studies also enables tutors to make objective assessments of student learning over a broad range of skills, a feature that can be more challenging with more traditional teaching methods.

Case Studies in Chemistry Teaching

Case studies have a long history in the teaching of many subject areas and their value with the subject area of chemistry has been documented during the past 20 years (*8,9*). Despite this, there are relatively few examples that have been reported within the literature for specific chemistry sub-disciplines. Whilst there are exceptions (*10-17*), these represent a surprisingly small number considering the reported benefits of case studies to students and their learning. Perhaps more unexpectedly, is the virtual absence of articles in the literature describing the use of case studies in analytical chemistry. Unexpected, since analytical chemistry has such a major contribution to the more 'real-life' applications of chemistry such as environmental or forensic science; areas that, in theory, should benefit greatly from the case study approach.

In contrast, there is clear evidence from pre-university chemistry teaching in the UK and the US that the context-based approach provides students with greater motivation to study chemistry at a time when numbers of undergraduate students choosing to study the subject at university are on the decline (*18-19*). Other studies have shown that some students perform better as a result of the context-based approach (*12, 18-20*) and it is tempting to make a link to this enhanced motivation. It is also clear that context-based teaching, including the use of case studies, is a growth area in terms of components or approaches to pre-university curricula. As such, increasingly, students are arriving at university with an expectation that this mode of teaching will feature significantly within their courses. A number of years ago, and at a time when the undergraduate chemistry courses in our respective institutions were undergoing substantial changes, we decided to dedicate time to developing a series of context-based case studies for use in undergraduate analytical chemistry programmes. Initially, the model was to create and embed material that would be particularly suitable for mid- and latter stages of the courses (Years 2-4 in UK BSc (Hons) and MChem courses). In practice, the researching and writing of this new material took approximately a year to complete, while the subsequent embedding into the course structures evolved over the following 2-3 years (see e.g. *17*) . In each of the Universities of Plymouth and Hull in the UK, these case studies are now employed as regular features of the analytical chemistry programmes across all stages of study and they have proven to be extremely popular with the students. As a result of this project's success, we have now written and piloted an additional series of problem-based case studies that interface between the secondary/tertiary levels (e.g. *21*). Although these further case studies have been developed with students on all chemistry courses in mind (and have been published by The Royal Society of Chemistry with this aim), they contain both cognate and discipline independent skills components that are particularly relevant to students studying analytical chemistry as a major component of their courses. The development of this second project has benefited greatly from our experiences gained from researching and writing more advanced level material previously.

Case Studies in Analytical Chemistry

The six case studies that we originally developed for analytical chemistry students are all set within a (fictional) town and region of the UK (Midshire). Each case study has analytical chemistry as its foundation, although the applications also include topics normally associated with pharmaceutical, environmental and industrial chemistry. The titles of the case studies together with their chemical themes are shown in Table I.

At the beginning of each case study, the specific scientific and transferable skills are presented to the students in more detail and these are further considered as the case study progresses. During this first session, particular attention is also paid to the assessment procedures, the wider expectations of the

Table I. Summary of Case Studies in Analytical Chemistry

Title	*Chemical and Other Themes*
A Dip in the Dribble	Analytical/environmental/industrial
Launch-a-Lab	Analytical/industrial/business
New Drugs for Old	Analytical/pharmaceutical
Tales of the Riverbank	Analytical/environmental
The Pale Horse	Analytical/forensic
The Titan Project	Analytical/industrial chemistry

students and the communication routes between the tutor and the students since these can be quite different to those experienced elsewhere in the course or previously. The case studies can be run in a number of different formats ranging from short 1-hour sessions to 3-hour (or longer) workshops. For all case studies, the students work in groups of 4-6 and it is the responsibility of the group members to organize any time needed outside of the workshop environment. In total, students are expected to dedicate in the region of 12 hours of study time to complete each study. Suggestions for management of the case studies according to these different formats are given in the comprehensive Tutor's Guides, which also provide overviews of the studies, assessment strategies, model solutions to some of the problems (or guidelines for more open-ended problems) and answers to frequently asked questions. These guides are available from the authors on request.

It is beyond the scope of this article to describe each case study in detail, but it is worth highlighting the features of two of the case studies; one that we have used mainly with students in their second year of undergraduate study (*New Drugs for Old*), a second, more elementary study that is more compatible with students working in their first year (*The Titan Project*). In addition, we describe how we have recently used the case studies in an on-line format for part-time students working in virtual groups. Details of some of the other case studies can be found elsewhere (*15-17, 21-22*).

New Drugs for Old

The case study entitled *New Drugs for Old* has been designed for students who have a moderate understanding of spectroscopic methods in analytical chemistry and the case study allows them to apply their skills within a pharmaceutical context. As such, *New Drugs for Old* is suitable for use with students in their second year of undergraduate study. During the course of the case study, the students need to devise a plan of action for both short and long-term investigation of a potential new analgesic drug isolated from a natural source, determine the structures of a series of extracted chemicals from selected spectroscopic data and physical properties, and propose a method for preparing the most pharmacologically active compound based on some suggested synthetic routes, economics and scale-up considerations. A summary of the general skills areas together with the specific activities is shown in Table II.

Table II. Summary of skills and activities encountered during the *New Drugs for Old* case study

Skills	*How skills are developed (Activities)*
Chemistry sub-disciplines	Analytical, toxicology, drug design, pharmacology, marketing
Scientific knowledge	NMR/IR/MS/UV-Vis data, synthetic schemes
Data treatment	Interpretation of spectra, quantitative analysis
Problem solving	Tackling unfamiliar scenarios, working with incomplete datasets
Communication	Oral presentations, report writing, end-of-session feedback
Improving learning	Use of feedback to improve individual and group performance
Information technology	Word processing reports, preparation of material for oral presentations
Planning and organization	Decision making, time management, prioritizing, working to tight deadlines

At the beginning of the case study, the groups are presented with a letter which has been sent to the Director of Research of a pharmaceutical company (Green-Chem Inc.) by an employee of the local university following an overseas trip to Malaysia. The employee, who is not a chemist, describes the analgesic effects of a plant extract that he was encouraged to take whilst suffering from symptoms including high fever. He offers a supply of the plant material to Green-Chem Inc. should this be of interest for further study. Having considered this background, the students then need to suggest a strategy for identifying the nature of the pharmacologically active component and to determine its potential as a new treatment for various symptoms. To complete the first task, the groups need to identify the steps that would need to be taken in order to produce an economically viable product. To achieve this, the students are given a series of

'Action Cards', which describe the individual parts of such a process and the task, effectively, becomes one of arranging these steps into an appropriate sequence. Amongst the things that emerge from this task is the need to identify (at an early stage) the chemical(s) that are responsible for the observed activity. This provides the rationale for the second task which involves the structural determination of three extracted and purified chemicals from a range of analytical (NMR/MS/UV-Vis/IR) and other physical data (M.Pt., CHN analysis, solubility, etc). Additional data for a known chemical is also available if it is felt that the students would benefit from some guidance in the first instance. Having completed the structural determinations, the most active component is identified (salicylic acid) and the groups evaluate some alternative methods for its synthesis, since extraction from the plant source is unlikely to be economically viable. Working with two simplified synthetic schemes, the economics for the preparation of salicylic acid are determined according to costs of reagents, reactions yields, scale-up considerations and any other factors that the groups consider important. Summaries and recommendations are presented by each group by way of written reports and/or oral presentations.

In the first session, where the groups are required to propose a project plan, the use of so-called Action Cards introduces the students to the numerous processes that are involved in pharmaceutical development according to their lowest common denominations (isolation, structure determination, synthesis, toxicity screening, formulations, approval, marketing, etc). What emerges from this activity is that project plans can be complex, do not necessarily have a strict linear sequence and, as such, there are not necessarily unique or 'best case' solutions. When presenting their plans, the students are reminded of the importance of providing a rationale for their proposals. In this way, students are prevented from making too many subjective proposals, though some students find it difficult to distinguish between solutions that are too general to be of any value from those that have a clear strategy, even if some uncertainties or unknowns remain. In contrast, most students are more comfortable with the subsequent session involving the structural determinations of organic chemicals since the methods and outcomes are more familiar or tangible. This confidence, however, is not always reflected in their success in identifying the correct assignments. In the final session, the problem solving elements of the previous activities are combined. For example, economic considerations for alternative synthetic schemes can be based on costs of chemical reagents from suppliers' catalogues and electronic sources, in combination with reaction efficiencies, etc. In contrast, the economics associated with scale-up, changing of reaction conditions, purchasing of intermediates (as alternatives to 'cheap' basic starting materials) can be more challenging to evaluate quantitatively. However, it is important for students to experience such different types of problem solving since few 'real life' scientific tasks reduce down to well defined algorithmic exercises. In this particular case study, the most logical suggestion from the student groups would be to recommend that Green-Chem Inc. do not pursue the development of the extract further since there are numerous world-wide market leaders representing the production of salicylic acid and its derivatives.

The Titan Project

The Titan Project can be used with students with a fairly basic knowledge of analytical chemistry and the case study allows them to extend their knowledge in an industrial context. It is suitable for use with students in their first or second year of undergraduate study. As a primary task, the students need to devise a plan for the future of an industrial site which gives them some insight into the factors which affect the work of industrial chemists, such as economics, safety and environmental legislation. They then consider the establishment of a new industrial analytical laboratory and review and select analytical methods. A summary of the general skills areas together with the specific activities is shown in Table III.

The Titan Project focuses on the industrial production of titanium dioxide (TiO_2) pigment. Titanium dioxide is the most widely used white pigment because of its high refractive index, brightness and opacity, and approximately 4.5 million tonnes are produced each year globally. The commercially useful pigment is produced in two forms, anatase and rutile, which have different physical properties and find applications in different industries. For example, rutile has high opacity and gives a high quality finish and is used in exterior paints, whereas anatase is more intensely white and is used in paper and ceramics. There are two industrial processes for producing the pigments; the chloride process, which produces only rutile, and the sulfate process, which can produce rutile or anatase. There are distinct advantages and disadvantages to each process. The chloride process is the more expensive as there are high energy costs and it utilizes expensive rutile ore as the raw material. The sulfate process produces enormous amounts of waste but uses cheap ilmenite ore as the raw material. Globally, the chloride process is increasingly dominant but sulfate plants still exist to cater for the anatase markets.

The case study begins with the students being given a presentation from a senior manager at the Titan Industries company. Titan Industries have recently purchased a titanium dioxide plant in Midshire and are planning to break into the European market. Historically, their business has relied exclusively on the chloride process but this new site contains a sulfate plant. The students are told that their task is to make recommendations for the future of the site. They must

Table III. Summary of skills and activities encountered during the *The Titan Project* case study

Skills	*How skills are developed (Activities)*
Chemistry sub-disciplines	Analytical, industrial, inorganic
Scientific knowledge	Pigments, synthetic routes, safety, environmental, analysis
Data treatment	Quantitative analysis, statistics
Problem solving	Tackling unfamiliar scenarios, working with incomplete datasets, using judgment, critical evaluation
Communication	Oral presentations, report writing, end-of-session feedback
Improving learning	Use of feedback to improve individual and group performance
Information technology	Word processing reports, preparation of material for oral presentations
Planning and organization	Decision making, time management, prioritizing, working to tight deadlines

produce a five year plan in which any expenditure is fully justified. They are provided with a briefing paper on the background to the two processes so that they can evaluate the chemical advantages and disadvantages of each. They are also provided with information about the site, the local area and the current market for TiO_2. They are presented with a map of the surrounding region which contains information about other industries and commercial activities, all of which can impinge on the students' deliberations. For example, a building materials manufacturer may purchase waste gypsum from the sulfate process and the local paperwork manufacturer may purchase the anatase it produces. Such regional interactions are of course artificial as the site would no doubt be part of a much wider marketplace, but it is successful in encouraging students to consider the range of interactions that affect industry. They are also given the front page from the local newspaper which includes stories related to local politics, the environment and other topical issues. The students have to consider all this information and produce fully justified plans after taking into consideration factors such as economics (raw materials, capital costs, energy consumption, manpower etc), the environment, safety, local politics, and the global market for the pigment. They present their plans to a representative from Titan Industries (the tutor). The sensible options that they propose include building a chloride plant whilst maintaining the sulfate process, building a chloride plant whilst closing the sulfate process, or maintaining the sulfate process. There is no single 'correct' answer to this activity and, as long as any recommendations are well argued and can be defended, they are accepted.

In the second stage of the case study, the students are informed that Titan Industries have decided to build a chloride plant on the Midshire site. This will involve setting up a number of new laboratories, one of which will be the environmental monitoring laboratory. This laboratory will monitor effluent from the plant for chloride ions. The students are given the task of reviewing suitable methods of analysis for chloride ions. An introductory activity encourages them to consider factors such as accuracy, precision, reliability, set up and consumable costs, levels of detection, speed and ease of analysis, etc. Each group of students presents a written report which reviews each available method and they identify their relative advantages and disadvantages. They are then required to consider the methods in more detail by carrying out statistical analysis on results for chloride ion analysis obtained by three different methods. They compare data gathered via titrimetry, gravimetry and colorimetry and discuss which of these methods they would employ based on their statistical analyses and other factors. The gravimetric analysis is undoubtedly the best for accuracy and precision, but time and technical demand would not make it a good choice for routine analysis in a busy laboratory. The activity finishes with a discussion of the merits of other analytical methods and the difference between carrying out analysis in an educational or research environment and in industry. The sensible choice for the latter, which the students can be led to, is an automatable, preferably on-line method, such as an ion-selective electrode.

On-line case studies

Previously, all of the case studies have been used in the classroom with students working in groups and this has proven to be highly successful (*17,21-22*). More recently, we have used some of the case studies with students who were studying at a distance and learning independently or, at least, with limited direct contact with their peers or tutors. In this application, the material has been used with a virtual learning environment (VLE). As far as possible, the way that the case studies are used mirrors the way that they are used in the classroom. The students are organized into groups and each group has its own communications area on the VLE where they can communicate via a discussion board and share files. The documents, handouts and slides that are used in the classroom version are placed on the VLE in the order that the students have access to the same information. The material can be made available all together or sequentially as the activity proceeds. Students are also provided with an outline of the case study, submission dates and a briefing on how the case study will run on-line. A general discussion board is established to answer any queries any student may submit. In this way, all students receive the same information and feedback. The students work on each task, using their group discussion board to communicate and share files. Submission of reports can be via hard copy in the usual way or through a digital drop-box on the VLE. One activity that cannot be reproduced from the classroom version is the oral presentation. This is replaced with on-line peer assessed Powerpoint presentations. Each group submits their set of slides via the general discussion board and then each individual student submits a completed peer assessment *pro forma* to the tutor. Peer feedback to each group can also be incorporated at this stage. When each stage of the case study is completed, the tutor can give feedback to all of the students via the discussion board. All the benefits of the case study approach, such as problem solving and team work are still present with the exception of oral communication. This is a small price to pay if it enables part-time or remote students to take part in a worthwhile learning experience. One further advantage of using the case studies on-line is that large cohorts of students can be handled relatively easily whereas large groups (> 30) are more challenging in a classroom.

Evaluation

Previously, we have described in some detail several of the impacts of using context-based case studies on student learning (*17, 21*). It is worth summarizing these points here, along with some further, more recent observations.

Firstly, and perhaps most importantly, end of course feedback demonstrates that students particularly welcome the opportunity to study chemistry within a real life context. This is consistent with the results obtained from a survey of 16-18 year old students taking science in schools in the UK (*23*) and no doubt, elsewhere. Students also like working with different problem types, especially as many other teaching styles and assessment procedures do not offer these. In particular, case studies enable students to work with problem types ranging from algorithmic exercises to open-ended problems which do not usually have a single or well defined correct solution (see e.g. *24* for a comprehensive description of problem types). For the latter, the end of session reflection periods are particularly valuable for the students, since many of them are not so familiar with the philosophies or boundaries associated with them, particularly during the early stages of their courses. Students also recognize how the case studies help to develop their transferable skills and commonly cite group work, time management and presentation skills as those areas of most significance for them. From a tutor's point-of-view, these are also areas where students gain in confidence having completed several case studies, and it is because of this type of skills progression that we endorse using cases studies throughout courses rather than as isolated activities.

Secondly, it is noteworthy that for those modules where the case studies are components, pass rates and coursework marks are usually at least as good as those obtained for other modules, and it is interesting to consider reasons for these. Apart from the observation that student motivation appears greater having studied via the context-based approach, we believe that case studies offer some further advantages for assessment over more traditional examination routes. For example, it is possible to test a greater range of skills through case study work and the outcomes are not so memory driven compared with examination based assessment. Further, since the students develop portfolios of information and data as the case studies progress, it is more feasible to assess for developments in the students' learning rather than evaluate snapshots that coincide with examination events. The format of the case studies (extended time period, student-tutor interaction, group work) also provides opportunities for other key skills to be assessed including critical thinking and judgment. For example, students can be asked to provide their rationale behind solutions to problems and engage in discussions with the tutor as part of the assessment, while the interactive nature of some of the case studies can have the effect of controlling their direction on an individual or group basis. Thus, those students/groups who present carefully considered and unambiguous requests for information from the tutor are rewarded with valuable feedback, which, in turn, usually results in higher grade achievement by the students (and *vice versa*). More prosaically, it is much more realistic to design and implement an assessment strategy that ensures that students achieve all of the necessary learning outcomes. Recent research suggests that there is growing evidence to indicate that this is often not the case for examinations even if it is claimed that they do so (*25*).

Finally, feedback from students on the use of on-line case studies has been very positive. Since 'distance learning' students rarely derive benefits from working with their peers, this cohort particularly appreciated the opportunity to collaborate with other students who are remote from their own location. These students also stated that the on-line case study made the independent learning experience less lonely. From the tutor's point-of-view, the quality of the assessed tasks produced via the on-line mode of learning has compared very favorably with the classroom based version. Apart from the satisfaction of delivering effective modes of teaching to all students, this type of observation is particularly important if it is expected that different groups of students achieve common module learning outcomes. Not surprisingly, the complexities of having to liaise with all group members and the tutor on-line in a meaningful and up-to-date manner required some support from the tutor. However, this was readily provided through the discussion boards, making this a flexible experience for staff as well as students.

Acknowledgements

We thank the many hundreds of undergraduate students at each of the universities in Plymouth and Hull who took part in the initial piloting of our case study material and to the Royal Society of Chemistry for funding the project through the Analytical Chemistry Trust Fund. We also thank those other teachers of analytical chemistry worldwide (>30 to date) who have requested copies of the case studies for use in their own teaching.

References

1. Dearing, R. *Skills for graduates in the 21st century*; The Association of Graduate Employers, Cambridge, 1995.
2. Mason, G. *Change and diversity: the challenges facing higher education;* The Royal Society of Chemistry, London, 1998.
3. Finer, E. *Chem. Brit.* **1996**, *32*, 3.
4. Fleming, J.; Sargent, M.; Singleton, V. *Chem. Brit.* **1994**, *30*, 29.
5. Woodget, B. Audit by the UKAP Skills Network of skills accumulation in undergraduate programmes of study. www.chemsoc.org/pdf/ukap/skills1.pdf (accessed Feb 4, 2002)
6. Subject benchmark statements. Chemistry. http://www.qaa.ac.uk/academicinfrastructure/benchmark/honours/chemistry.asp (accessed Feb 4, 2002).
7. Guidelines for preparing programme specifications. http://www.qaa.ac.uk/academicinfrastructure/programSpec/progspec0600.pdf (accessed Feb 4, 2002).
8. Garratt C. J.; Mattinson B. J. H. *Education Industry and Technology,* Pergamon Press. 1987.
9. Pontin, J. A.; Arico, E.; Pitoscio Filo J.; Tiedeman P. W.; Isuyana R.; Fettis G. C. *J. Chem. Educ.* **1993**, *72*, 525.
10. Bennett, N.; Cornely, K. *J. Chem. Educ.* **2001**, *78*, 759.
11. Breslin, V. T.; Sanudo-Wilhelmy, S. A. *J. Chem. Educ.* **2001**, *78*, 1647.
12. Holman, J.; Pilling, G., *J. Chem. Educ.* **2004**, *81*, 373.
13. Cornely, K. *U. Chem. Ed.,* **1998**, *75*, 475.

14. Cheng, V. K. W., *J. Chem. Educ.* **1995**, *72*, 525.
15. Belt, S. T.; Phipps, L. E.; *U. Chem. Ed.* **1998**, *2*, 16.
16. Belt, S. T.; Clarke, M. J.; Phipps, L. E. *U. Chem. Ed.* **1999**, *3*, 52.
17. Belt, S. T.; Evans, E .H.; McCreedy, T.; Overton, T. L.; Summerfield, S. *U. Chem. Ed.* **2002**, *6*, 65.
18. Burton, W. G.; Holman, J. S.; Pilling, G. M.; Waddington, D. J. *J. Chem. Educ.* **1995**, *72*, 227.
19. Gutwill-Wise, J. P. *J. Chem. Educ.* **2001**, *78*, 684.
20. Sutman, F. X.; Bruce, M. H. *J. Chem. Educ.* **1992**, *69*, 564.
21. Belt, S. T.; Leisvik, M. J.; Hyde, A. J.; Overton, T. L. *Chem. Ed. Res. Prac.* **2005**, *6*, 166.
22. Summerfield, S.; Overton, T.; Belt, S., *Anal. Chem.* 2003, **75**, 181A.
23. Student review of the science curriculum. Major findings. http://www.planet-science.com/sciteach/review/Findings.pdf (accessed Dec 7, 2003)
24. Johnstone, A. H. *U. Chem. Ed.* **2001**, 5, 69.
25. Bennett, S. W. *U. Chem. Ed.* **2004**, *8*, 52.

Chapter 8

Service-Learning: An Oxymoron in the Physical Sciences?

Or: 10 years of Service Learning: Where to?

A. Fitch

Department of Chemistry, Loyola University at Chicago, 1068 West Sheridan Road, Chicago, IL 60626

An analysis of peer-reviewed articles, abstracts, and presentations with the words "service-learning" and chemistry over the last ten years reveals some interesting trends in the field. The data analysis suggests certain commonalities and points to pitfalls as well as areas where resources are lacking.

This article examines the past ten years of activity in the area of service learning in chemistry. The data are categorized and used to draw conclusions about the future guises that service learning may wear in chemistry.

Several search engines were employed to find examples of service learning in chemistry. Google Scholar unfiltered returned approximately 48,000 hits for the Service Learning in Chemistry. An advanced search with the exact phrase "Service Learning" and request for Chemistry returned 11 hits. Of those 11 hits, one was for bus "service" at a chemistry learning conference and one for the Park Service opening sites for students. Two listed C. K. Larive describing the utility of the Analytical Sciences Digital Library (ASDLib. http://www.asdlib.org/index.php (December 7, 2006)) in finding sources for service learning via the web, citing my own professional web page (Alanah Fitch. http://www.luc.edu/faculty/afitch/Service%20Learning.htm. (December 7, 2006)) Only two peer reviewed publications were culled. The first was an article by myself and colleagues with biographical sketches where my bio brings up the word service-learning. The second is one of a small sub-section of citations found by SciFinder Scholar.

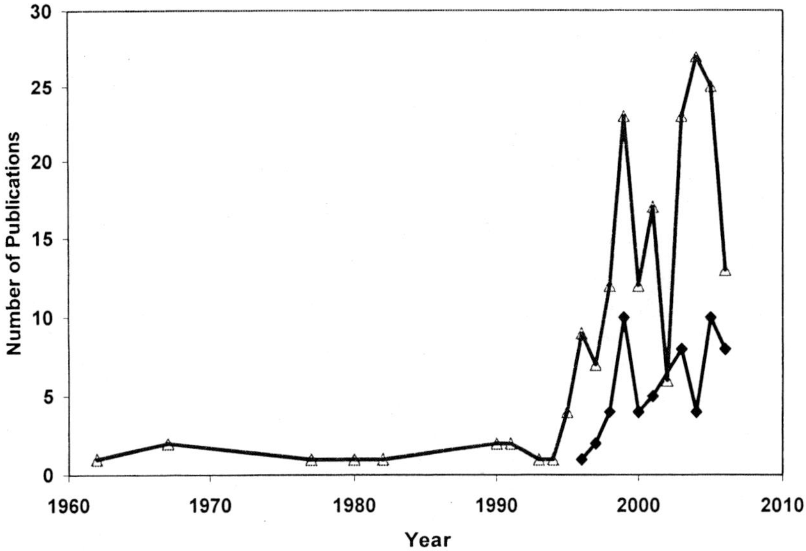

Figure 1. Number of citations for Service Learning from SciFinder as a function of publication date. The upper line is the total number of citations and the bottom line has been further filtered "by hand" for purely chemistry and chemically related articles. The discrepancy is primarily from the range of nursing articles.

The SciFinder data base was queried separate times with permutations on Service Learning and Chemistry until it appeared that all possible hits were drawn. A general search via SciFinder for references to "service-learning" leads to 2443 hits. These hits were refined by filtering for "chemistry". Depending upon the exact sequence of search one can obtain slightly different data bases so the query was repeated until no new hits emerged for service learning and chemistry. This data base was 191. An examination of the abstracts revealed that a large segment came from the field of nursing. Indeed, all publications prior to 1996 were not in the direct field of chemistry (Figure 1). Filtering out contributions from the field of nursing and random articles culled from neurochemistry and neuropsychology on left a total of 56 citations.

A third search made use of the Analytical Sciences Digital Library. This library contains already sorted and peer reviewed sources related to analytical chemistry and instrumentation. Two sources of information were culled from the ASDLib. One is a website from the University of Notre Dame for their Chemistry 331class (Dennis Jacobs. Chemistry in Service of the Community.

http://www.nd.edu/~djacobs/chem331.html (December 7, 2006)). This describes a class developed in a multidisciplinary fashion from the departments of Chemistry and Biochemistry, Civil Engineering, and Geological Sciences, along with several major community stakeholders. The class provides results on mold and lead directly to households in the South Bend community. The second "hit" is an article by Deborah Wiegand (University of Washington) and Melissa Straight (Alma College) (*1*). This article summarizes the results of a Service Learning Blueprint meeting of seven chemistry faculty incorporating service learning into their curriculum in 2002. Following the citations in that article culled two more articles (*2-3*) one of which appeared in the other searches. Excluding web sites the total culled citations are 58.

These 58 citations represent a maximum of 6 references per year over 10 years which does not seem to imply a rapidly growing or expanding field. It should be immediately acknowledged, however, that the list of publications may be vastly under reporting "service-learning activities. For example, 1996 is the date of my own first publication, an A-page *Analytical Chemistry* description of our single analyte lab on lead (*4*). It, like many others, does not show up in a search for "service-learning".

Figure 1 plots the growth in references as a function of date. There appears to be a "take-off" point around 1996 with an initial burst of energy which slumps around 2001 and a restart in 2002. This would be consistent with the introduction of the "broader impacts" requirement instituted by the National Science Foundation at the behest of Congress in 2002 (The Science House. Broader Impacts. http://www.science-house.org/visitor/impacts.htm (December 7, 2006)).

Even more interesting is the fact that of the 58 citations only 9 are peer reviewed publications (*1-3, 5-10*). This leads to the interesting supposition that "service-learning" in chemistry *really* is an oxymoron. The etymology of the word oxymoron shows two Greek roots. "Oxy" derives from "sharp or keen" and "moros" from "foolish". A classic example of an oxymoron is "deafening silence", a phrase which can certainly be used to describe the past 10 years of service-learning in chemistry.

Why are there so few citations and even fewer peer reviewed article in service-learning in chemistry? Let us begin by asking *Who* is doing service-learning in chemistry and *What* is described as service learning. Perhaps the small number of citations is related to the fact that the time required for success in competitive grant activity is so high that service learning is simply beyond the accessible time for large research oriented institutions. Taking the 56 citations from SciFinder, listing all authors (379) and sorting by institution leads to Figure 2a. This data would suggest that there is no particular bias in type of institution. A good case can be made, for example, that some service learning activities may be better supported at institutions with greater resources or with faculty dedicated to the field of chemical education. Sorting the same 379 individuals by gender where it was easily ascertained leads to Figure 2b.

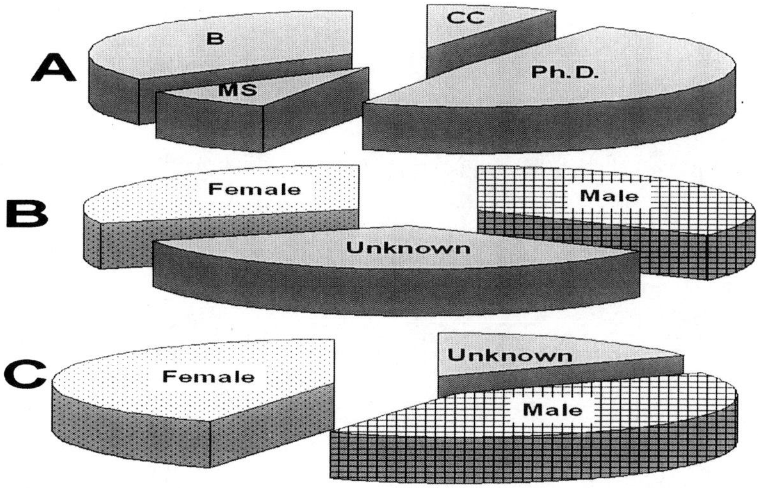

*Figure 2. **A.** Distribution of authors by type of institution where B is primarily UG, CC is a community college. **B**. Distribution of all authors by gender. **C.** Distribution of authors with multiple citations by gender. In "B" and C" solid gray represents authors where gender was not clear.*

117 authors were female, 131 were male and 131 were of unknown gender. Assuming that many of the authors were one time only contributors during their graduate career, individuals who were authors of multiple citations (42) were sorted by gender resulting in 18 female and 17 male (Figure 2c). The interest in service-learning is relatively equivalent by gender as judged by publication. On the other hand, it should be pointed out that the pool of university chemistry professors is strongly biased toward men and the publications record, if gender neutral, would be expected to be similarly biased.

What types of activities fall under the rubric of "service-learning"? A search for the definition of service pulls up a number of entries. Service is "an act of helpful activity; help; aid: to do someone a service". I define two types of "service-learning" activities to fall under this label. One consists of researching, making, doing, and performing demonstrations for the public, K12 institutions, or museums in the area of chemistry. This model represents mapping of literacy training activities from the humanities into the sciences. An example of a peer reviewed article in this area is that of Esson *et al.* (*7*).

A second type of activities are those that I define as better categorized as "*active-learning* projects in a social context with a very loose connection to service". Examples of this latter category are general environmental surveys that are neither elicited by the community or under any particular time deadline for a policy decision. They may also consist of surveys the data of which will never enter into a peer reviewed data base. These activities are those in which the coaches should spend some amount of time thinking about the ethical dimensions. As an example, if students are mapping for posited downstream pollution from farms, are the students engaging in a form of "scare science"? The activity may be justly engaging for the students and raise the learning level, but what impact does it have for the community which is or can be affected by the activity?

There is a second set of definitions to service which should be considered. These definitions all gravitate toward the transfer of technical knowledge in a value added transaction. For example "the supplying or supplier of utilities or commodities, as water, electricity, or gas, required or demanded by the public". Figure 3 shows the chronological citations using these three definitions. A fourth category was created to track articles whose subject matter was the utility of "service-learning" on learning and/or the mechanics of "how to do" service learning. As shown there is no particular trend in any of the categories with time. The running total of citations is biased strongly toward the first two categories of "service". 19 publications fell in the category of "science-literacy"; 10 in the category of "active-learning in a social context"; 11 in the "how to" category and only 7 in the "on demand" service.

What are the models of successful "Service as a Technological Deliverable"? Two examples will be given below, each representing distinct strategies in providing a "technological deliverable". Each must address certain constraints which I have called a "service-learning research matrix" as shown in Figure 4. For research data to be validated and useful the topic is generally tightly controlled and defined in a broader context of the discipline. Sampling and analysis follow written protocols and quality assurance is built in. For watershed sampling to move beyond "active-learning" and into "service-learning" the outcome should be subject to the same controls as would occur in a traditional research model described above. Figure 4 shows how a watershed sampling may result in true *service* while students are still learning the tools of the trade. Here there is an agency which has a temporal life span sufficiently long enough to outlive the span of a single class and which has a vested interest in the results so that quality assurance of the results is ongoing. The coach of the class specializes in a targeted few technologies which can be set up to minimize quality control issues. The downside of such a structure is that some of the problem solving skills that one would like the students to acquire may be removed from their sphere. Defining the exact nature of the problem, the methodology used to acquire the data, and the procedures used for quality assurance may be set by the coach and the sponsoring agency. An example of an agency/class marriage is the work carried out by the classes of Anna Cavinato at Eastern Oregon University (*11*).

As we pointed out in an earlier article on the ethics of service learning (*12*) "Service as a Technological Deliverable" suffers from mismatches between the "client" and the classroom. J. Q. Public has been educated by the mass media to

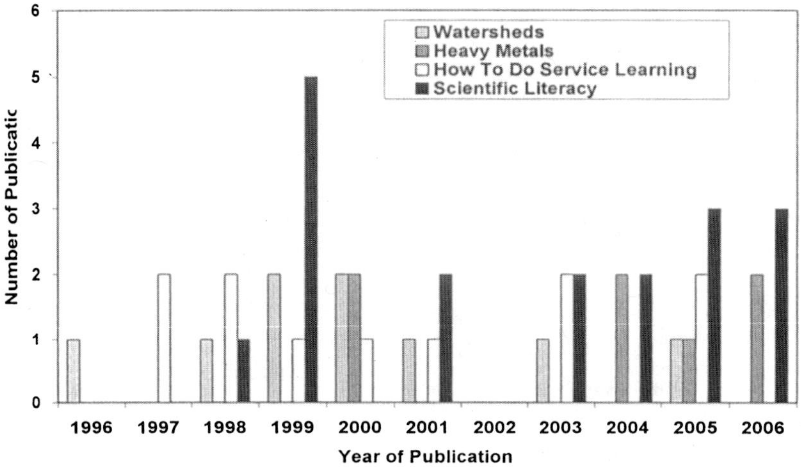

Figure 3. Type of service activity in publications as a function of year of publication. Solid gray represents "science literacy" activities. Dotted represents activities better defined as "active-learning in a social context". Black represents lead and arsenic analysis with data reported to the public. The final category (diagonal lines) represents "how to do" service learning articles.

expect an instantaneous and "true" (whatever that may be) result from scientists and their apparatuses. They may be impelled to seek help by some immediate medical or political (policy decision) need. Data in the academic laboratory, on the other hand, is often generated rather slowly both due to competing time constraints (particularly acute for the undergraduate) and level of training. As a result "on demand" service is difficult to do and, thus, not surprisingly there are few reports of "on demand" service-learning classes. One exception to this rule is the report by Joseph Gardella at SUNY Buffalo of a class that is responding to queries posted to an institutional portal for services (*13*). This model parallels the EU system of science shops (Living Knowledge, The International Science Shop Network. http://www.scienceshops.org/ (Accessed December 7, 2006)) which distributes J. Q. Public requests to a variety of institutions. The vast majority of such requests while "science" are not in the physical sciences and consist of social and economic science activities.

The Service-Learning Research Matrix in this case requires a mediator to frame the question between J. Q. Public and the coach/classroom. Either the coach already has the skills for the validation process or J. Q. Public has been vetted to expect non-validated results and/or results which are accrued for a longer time frame than he/she may have initially expected. This model has been

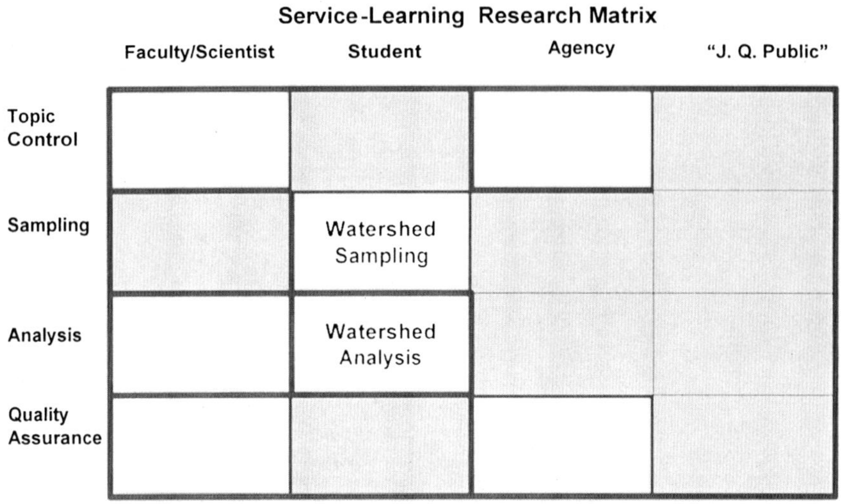

Figure 4. *A service-learning research matrix for watershed analysis. The bold boxes represent areas in which varies stakeholders participate. Here the faculty member, in conjunction with a governmental agency, has determined what constitutes desirable data. The faculty member with the agency is providing quality assurance. The sampling and data analysis are carried about by the students. The resulting data, because of tight control of topic and quality assurance, may end up in a useful public policy data base*

enthusiastically promoted by Loyola's Center for Urban Research and Learning. The primary difficulty is that within any given institution there are a finite number of faculty with their associated classes - each chemistry department could conceivable mount one perhaps a maximum of two vetted classes. One could image that years might go by before Mr./Ms. Public call upon a class to solve a particular problem within the expertise of the coach.

Other issues that arise for this second model occur at the attribution and ownership level of the data thus obtained. If publishable, how many student authors will appear? Does it matter to the reviewers? The work of three semesters of research into heavy metal contamination of soils near the North West Municipal Waste Incinerator by Instrumental Methods of Analysis Chemistry students at Loyola University hit significant difficulty in the review process because of concerns about quality control and number of authors. Does the data belong to J.Q. Public, even though all of the data analysis and quality control lie in the hands of the university?

Another model is possible - in which all students in all classrooms over several years participate in a targeted analysis (similar to watershed above). The aggregate data could be loaded to a single data base with a statistical meta analysis of that data. There are many good examples of similar aggregated scientific data bases. These examples include NMR, MS, and X-ray data bases. Some of these are for profit assemblages (e.g. Wiley's M.S. library) and others are "public" to a group of users sufficiently educated to utilize and expand the resources. How to organize the data and perform an automated internal statistical meta analysis becomes the main issue in the utility of the data base. Some of these questions are currently being explored but not, generally, in the context of creating usable student driven data bases for public policy decisions (*14*).

To summarize, literature on "service-learning" in chemistry began appearing about 10 years ago and appears to have been given a slight boost by "broader impact" requirements in grant proposals to the National Science Foundation. Most of the examples surveyed consist of "active-learning in a social context" or of examples of "scientific literacy". The literature culled from several usually reliable resources suggest that few of the service-learning activities result in peer reviewed publications either as an exemplar of teaching, or as a source of validated data useful for policy analysis. In order to overcome this latter barrier tools are required by which aggregated data from multiple classes are subjected to a statistical analysis for quality control. On the gloomy side, based on the lack of "deliverables", the future of the field appears dim. On the sunny side, the vast majority of articles that fell within the search parameters suggest that there are significant benefits to students whose educational experience encompasses some form of service learning.

References

1. Weigand, D.; Strait, M. *J. Chem. Educ.* **2000**, *77*, 1538-1539.
2. Kesner, L.; Eyring, E. M. *J. Chem. Educ.* **1999**, *76*, 920-923.
3. Shachter, A. M.; Edgerly, J. S. *J. Chem. Educ.* **1999**, *76*, 1667-1670.
4. Fitch, A.; Wang, Y.; Mellican, S.; Macha, S. *Anal. Chem.* **1996**, 727A-731A.
5. Landolt, R. G. *J. Chem. Educ.* **2006**, *83*, 334-335.
6. Chauhan, A. *Chem. Eng. Ed.* **2005**, *39*, 296-300.
7. Esson, J. M. *J. Chem. Educ*, **2005**, *82*, 1168-1173.
8. Grover, N. *Biochem. Mol. Bio. Ed.* **2004**, *32*, 367-372.
9. Draper, A. J. *J. Chem. Educ.* **2004**, *81*, 221-224.
10. Cracolice, M.; Ward, K. *Chem. Educator*, **1998**, *3*, 1-11.
11 Cavinato, A. G. *Abstracts of Papers,* 230th ACS National Meeting, Washington, D.C, Aug. 28 – Sept. 1, 2005; American Chemical Society: Washington, DC, 2005; CHED-113.
12. Fitch, A.; Reppman, A.; Schmidt, J. In *Acting Locally: Concepts and Models for Service-Learning in Environmental Studies;* Ward, H., Ed.; American Association of Higher Education: Washington, DC, 1999, 30-37.

13 Gardella, J. A. *Abstracts of Papers*, 221st ACS National Meeting, San Diego, CA, April 1 – 5, 2001; American Chemical Society: Washington, DC; CHED 948.

14. Frenklach, M.; Packard, A.; Djurisic, Z. M.; Feeley, R.; Russi, T.; Golden, D. M.; Gupta, A.; Bowman, C.T.; Green, W. H. Jr. McRae, G. J.; Smith, P. J.; Wang, H.; Westmoreland, P. R.; Allison, T. C.; Frey, J. G.; Pilling, M. J. *Abstracts of Papers,* 232nd ACS National Meeting, San Francisco, CA, Sept. 10 – 1, 2006; American Chemical Society: Washington, DC, 2006; PHYS 277.

Chapter 9

Service Learning in Analytical Chemistry: Extending the Laboratory into the Community

A. G. Cavinato

Department of Chemistry and Biochemistry, Eastern Oregon University, One University Boulevard, La Grande, OR 97850

Service-learning provides an opportunity to actively involve students in projects of analytical nature. In this article continuing efforts at incorporating service-learning projects into the laboratory curriculum are decribed with the specific goal of engaging students in the analytical process. The first example describes an on-going project with a local organization, the Grande Ronde Model Watershed, where students compile a database of water quality parameters to assess the impact of restoration on the quality of water at End Creek. The second example describes a partnership with scientists from the Confederated Tribes of the Umatilla Indian Reservation to assist the Tribe in addressing environmental issues of interest and foster joint research projects between students at Eastern Oregon University and Tribal members.

Many educators recognize the pedagogical value of experiential learning where students can apply skills learned in the classroom to the solution of real-world problems. Service-learning, in particular, is increasingly being recognized as a valuable tool to actively engage students in learning. In its broader definition, service-learning includes a wide array of experiential endeavors from volunteer and community service projects to field studies and internship programs. In community-based service-learning, students work to address a significant real world issue in responsible ways, while developing an improved sense of commonality with the community partner (*1*). Jacoby defines this pedagogy as "a form of experiential education in which students engage in activities that address human and community needs together with structured opportunities intentionally designed to promote students learning and development" (*2*). This pedagogical approach seems particularly suited to address concerns in preparing students for future careers in sciences where both technical and non technical skills are required along with the ability to think critically, communicate effectively, and work as a helpful team member (*3*). The American Chemical Society, through its Experiential Programs in Chemistry, encourages chemistry departments to include service-learning as an important component in undergraduate chemical education (*4-5*). The ACS Committee on Professional Training (CPT) in its proposed revisions of the guidelines for approval of bachelor's degree programs in chemistry stresses that in addition to "educating and training students in chemical concepts and practice", excellent chemistry programs also "produce students who work safely in the laboratory, demonstrate effective oral and written communication, and work effectively as a member of a team." The recommendations also state that "excellent programs encourage students to ask questions, design experiments, interpret results based on current scientific information, exhibit ethical scientific conduct, and develop behaviors and thought patterns leading to innovation and a capacity for lifelong learning" (*6*).

Service-learning may help students achieve some of these important skills (*7-9*). Ward comments that students engaged in service-learning projects may develop skills that place them at an advantage for future employment (*7*). If graduates have the ability to operate in a multi disciplinary environment where success of the project relies on the shared expertise of group members, have the confidence and motivation to solve a given problem and the ability to communicate the results of their work, they are at a clear advantage when seeking employment. Ward also identifies positive outcomes for teachers, institutions and community partners. Faculty benefit from working with more engaged students and may find that involvement with community partners brings new opportunities for professional development and networking. Colleges and universities also benefit from improved relationships with the community. Service-learning projects may be a way for university administrations to lessen tensions with the hosting community and demonstrate a desire to contribute to the improvement of the community welfare.

Numerous articles describe the implementation of service-learning projects in chemistry (*10-12*), with several highlighting specific examples of analytical and environmental applications (*13-19*). For example, students assist community groups in measuring lead content in samples from nearby neighborhoods (*12-13*); analyze ambient and indoor air for a variety of pollutants including volatile organic compounds (VOC), particulate matter and mercury (*15, 18*); or monitor water quality of local streams (*14, 16, 18*). The National Service-Learning Clearinghouse, Chemistry resource site (*20*), and the Analytical Science Digital Library (*21*) offer excellent on-line resources for educators interested to learn more about service-learning.

In this article, two recent service-learning projects developed at Eastern Oregon University are overviewed. The first example describes an on-going project with a community organization, the Grande Ronde Model Watershed, where students compile a database of water quality parameters over a period of ten years to assess the impact of restoration on a local stream. The second example describes a partnership with scientists from the Confederated Tribes of the Umatilla Indian Reservation to assist the Tribe in addressing environmental issues of interest and foster joint research projects between students and Tribal members. Details of both projects along with assessment strategies, successes and challenges encountered are discussed.

Service-learning projects at Eastern Oregon University

"The service-learning experience really opened my eyes to what can be done to our surrounding environment with the knowledge that we are gathering while in school. Before what I was learning in school seemed to have very little real life application, but this project really showed that it is not the case"

Chem 206 student – Spring '06

The implementation of service-learning projects in the chemistry curriculum at Eastern Oregon University was in part influenced by the existing campus culture that highly values experiential learning as part of the undergraduate educational experience. Although not yet a graduation requirement, students can complete the four "Cornerstones of Learning" which include at least one-quarter of independent research, an internship in an academic or industrial setting, an international experience and the completion of a community-based project. I had for quite some time contemplated the idea of incorporating more real-world projects in my courses and this opportunity materialized when I was approached by two different community partners.

The End Creek project: Monitoring and assessing water quality in the general chemistry laboratory

This service-learning project was first implemented in spring 2005 as part of the laboratory portion for the third term of the general chemistry sequence for science majors. Most students is this course are freshmen with limited laboratory skills. Typical enrollment is 60 students divided into three laboratory sections. The course includes 3 x three-hour lectures and six hours of laboratory per week and provides the first introduction to chemical equilibria, including acid/base and solubility equilibria and some elements of electrochemistry. For many years, the laboratory component focused on traditional qualitative analysis involving the identification of cations and anions through wet chemistry protocols. While the qualitative analysis laboratory contributes to the development of good laboratory skills and the identification schemes challenge students, there were many reasons for change. As I began shaping the new curriculum, I was guided by the following learning goals and objectives:

- Provide students with a research like experience
- Incorporate more analytical and instrumental aspects earlier on in the curriculum
- Challenge students with real-world problems
- Improve oral presentation and writing skills
- Gain a better understanding of regulatory policies and agencies
- Provide opportunities to become involved and invested in the community

The opportunity for change presented itself when I was approached by staff members of the Grande Ronde Model Watershed (GRMW), a non-profit organization whose mission is to enhance the local natural resources. GRMW scientists were actively seeking potential collaborations with the university as a means to supplement their monitoring capabilities limited by scarce resources and lack of equipment. Particularly, they were interested in assessing the impact of stream restoration on the quality of water at End Creek. To accommodate crop production, the creek was straightened and confined in the 1930's resulting in diminished fish habitat. The goal is to transform the current straight and habitat deficient creek into a sinuous habitat-rich channel with access to its seasonal flood plane.

In the spring of 2005, the first general chemistry class started a ten-year sampling effort to document pre-construction conditions and allow the establishment of a "base line" set of data to be compared with data collected during and after project implementation. In documenting potential water quality change as a result of the project, students participate in an actual conservation project providing experience outside the laboratory and classroom environment. Specifically, GRMW requested sampling of nitrates, phosphates, dissolved oxygen, pH, and total suspended solids, all water quality parameters of concern in the Grande Ronde Basin. Because I wanted students to gain familiarity with additional methods, I added a qualitative assessment of metals by atomic absorption spectrometry, volatile organic compounds (VOC) by purge-and-trap GC-MS, and most recently the analysis of coliform bacteria by standard microbiological techniques. Students sampled three sites in the project area: one above, one in the middle, and one at the bottom of the project, allowing us to compare water quality throughout the project as well as to identify any changes that might occur in water chemistry as the stream flows through the project site. A schedule of the laboratory activities is outlined in Table I.

Table I. Example of laboratory schedule

Week	*Task*
Week 1	Orientation – Service-learning project Meet GRMWS representatives
Week 2 - 3	Qualitative Analysis – selected cations/anions
Week 4	Sampling with GRMWS scientists Begin laboratory analyses
Week 5 - 9	Analyses continued
Week 10	Analyses report due (labnotebook) Poster/slide presentation draft
Final	Final presentation to GRMWS Reflection paper due

During the first week of the quarter, scientists from the GRMW visit the classroom and introduce the project. They provide a short history of the creek and the rationale for the restoration project. The purpose of this introduction is not only to provide specific information about the project in terms of the current riparian conditions and goals for restorations, but also to raise students' awareness of environmental policies and the role that many different organizations play in the implementation of a restoration project of such dimension.

In subsequent classroom meetings, students are incrementally introduced to the different parameters they will be measuring and provided with background information about the instrumental techniques they will be using in the laboratory. Students are directed to "Volunteer Stream Monitoring" (*22*), a very valuable website maintained by the EPA, which provides background information about the analytes of interest as well as appropriate protocols for water sampling and a field sampling data sheet. Detailed experimental procedures are provided at laboratory time and are also made available in the course website.

During the fourth week of the term students conduct sampling at the site under supervision of the GRMW scientists. All measurements performed as part of the project, including on-site and off-site, are listed in Table II. Appropriate sampling techniques as recommended by EPA are enforced, including use of acid-washed containers and gloves. Choice of sensors and techniques is geared at demonstrating differences in the sensitivity and detection limits. For example, nitrates and phosphates concentrations so far detected at the creek in all three sampling sites fall below the detection limit of the Hach kits. However, when the same analyses are conducted in the laboratory with more sophisticated methodologies, presence of small amounts of nitrates and phosphates become apparent.

Table II. Measurements performed as part of the project

On-site measurements	*Off-site measurements*
Temperature	Nitrates (ion selective electrode)
pH (Hach kit and solid state sensor)	Total orthophosphate (ascorbic acid/spectrophotometric)
Dissolved Oxygen (Hach kit and electrochemical)	Total Suspended Solids
Nitrates (Hach kit)	Metals (AAS)
Phosphates (Hach kit)	VOCs (GC-MS)
	Coliform analysis

At the completion of on-site analysis, a considerable amount of water is sampled and stored in acid washed polyethylene containers kept on ice and brought back to the laboratory for subsequent analysis.

Back in the laboratory students are divided into small groups and assigned on a rotation basis a specific experiment on the water sample from the group's specific sampling site. If time permits, students analyze more than one sampling site.

Alternative methods to quantitatively measure nitrates and phosphates are chosen to introduce students to concepts of instrument calibration (*23*). Students make up reagents and prepare all standards, samples and reagent blank. Each measurement is the average of three independent analyses conducted on the water of a given sampling site.

Total suspended solids (TSS) are measuredby mass difference (*23*). The experiment, though very simple, does challenge students with the use of the analytical balance. Because the average TTS are very small, it is not unusual for students to obtain a negative values. This discrepancy allows for discussion of appropriate handling techniques for analytical mass measurements.

Hands-on exposure to more advanced instrumental techniques is accomplished by using flame atomic absorption spectroscopy for a qualitative estimation of presence of metals in water (*24*) and purge-and-trap GC-MS for detection of VOCs (*25*). For the AAS experiment, students are reminded of familiar concepts of absorption from conventional spectrophotometry and then introduced to the operation of the flame atomic absorption spectrometer. Water is checked for presence of iron, zinc, copper, chromium, nickel and lead. None of these metals have been detected so far, except for very small traces of iron. A brief introduction to the use of GC-MS for environmental monitoring is also provided and students are asked to research possible sources and the significance of the presence of VOCs in the environment. In both cases, students have hands-on access to the instrumentation under supervision of TAs.

In spring 2006, an additional experiment dealing with the determination of coliform bacteria was added (*26*). The decision was prompted by the community partner's interest in potentially identifying presence of coliform bacteria in the creek, and to make students aware that coliform analysis is a standard water test often required to certify water usage. The experiment also bridges across analytical and microbiological techniques, thus showing the relevance of good laboratory practice across disciplines.

During the first three weeks of the project students are given the opportunity to receive feedback on their labnotebooks' organization and calculations and make revisions before a grade is assigned. This strategy helps students improve organization (tables, graphs, etc.) and ensures better quality control of the data. At the completion of all experiments, students are given one more week to complete all calculations and reports and draft a slide presentation of the results. The final presentation to the community partner is scheduled during finals' week. To avoid repetition, each group is asked to focus the presentation on one designated analyte.

Service-learning in Environmental Chemistry: A partnership with the Confederated Tribes of the Umatilla Indian Reservation.

A second model of service-learning was implemented in an Environmental Chemistry Laboratory Course as a partnership with scientists from the Confederated Tribes of the Umatilla Indian Reservation (CTUIR). Students in this upper division laboratory, being more advanced, are expected to independently design experiments that can help answer questions regarding environmental issues facing the Tribe. Unlike the GRMW project where students receive protocols and are much more guided in the procedures, students in the environmental project have to device their own experimental design and implement appropriate methods of analysis with limited guidance.

The goal of the partnership is to enable students and tribal members to engage in joint research projects. Students are given the choice of pursuing a specific project while Tribal members are given access to expertise and instrumentation that can provide critical data and information on environmental issues impacting the reservation and ceded lands.

Prior to the beginning of the course, CTUIR scientists and EOU faculty meet to identify projects of common interests that have the potential to provide meaningful data within the timeframe of the course and with available equipment and instrumentation. Examples of projects are shown in Table III, inclsing those that proved non-viable.

The course follows a calendar similar to the one presented in Table I. The course meets for three hours every week and has an enrollment of five to six students.

Table III. Proposed projects in the last five years

Project title	*Rationale*
Analysis of a pond's sediments for lead	Potential hazardous waste disposal site
Investigation of the snow pack composition of a commercial area within the reservation	Impact of cars and recreational activities on the quality of the water supply
Analysis of ceremonial mosses for potential presence of PCBs , PHAs, metals	Potential contamination of native plant species used in traditional practices
Indoor air quality in local recreational facility	Exposure to tobacco smoke at a gambling facility
Analysis of salmon liver from the Columbia River	Not pursued due to excessive costs and complex procedures

On the first day of class, students are introduced to the concept of service-learning and have an opportunity to discuss project ideas with scientists from CTUIR. During the second week of class, students divide into small groups and, based on the information that they gather from the literature search, choose a specific project. At the beginning of the third week, students provide a plan outlining the specific method of analysis to be employed, the sampling strategy, a list of chemicals and equipment needed, and the safety hazards potentially posed by the chosen laboratory procedures. The plan is reviewed by at least one faculty member and further discussed with a CTUIR scientist to ensure that it can be carried out. Sampling is conducted at different CTUIR sites during the fourth week of the term with assistance from CTUIR scientists. Students are introduced to fundamental statistics of sampling (*27*) and are encouraged to consider factors such as sample size and number of replicate analyses. For example, in the case of the snow pack composition project, CTUIR scientists directed students to dig a 2 meter pit in the snow and sample at constant intervals from the surface to the bottom of the pit. The next four weeks are spent in the laboratory conducting the actual analyses. Students quickly gain an appreciation for the challenges posed by real-world samples. For example, they have to learn more advanced calibration procedures for the determination of lead in sediments by graphite-furnace atomic absorption spectroscopy (*28*) and VOCs in snow samples (*25*). In some cases the complexity of the analyses involved did not allow for completion of the project, although information gathered by students in a previous year was subsequently used by a new group of students in the following year. The final presentation to the CTUIR scientists consists of either a poster or slide presentation summarizing the results of the project. Several students have since wanted to continue some of these projects as independent research projects and have presented the results of their finding at professional meetings.

Assessment strategies

Multiple tools were used to assess the impact of the service-learning experience on students' learning. These include individual laboratory reports, the final presentation to the community partner, and a reflection paper on the overall experience. In addition, in the case of the water monitoring project within the general chemistry course, a pre-test/post-test assessment was also administered.

The pre-test/post-test quiz consisting of fifteen multiple choice questions was designed to assess whether the service-learning experience had any impact on students' knowledge and understanding of specific analytical concepts and more general environmental issues. Specifically, questions 1-5 probed students' knowledge of environmental regulations and agencies (ie. local sources of contamination, EPA, DEQ, maximum contaminant level, etc.); questions 6-10 addressed analytical techniques (sampling, preparation of standards, dilutions, calibration curves); and questions 11-15 assessed students' understanding of instrumental parameters (detection limit, sensitivity) and specific instrumental techniques (AA and GC-MS). The quiz was administered at the beginning of the term prior to any discussion of the service-learning project and again at the end of the term once all laboratory experiments and data analyses were completed.

The assessment conducted over the past two years indicates increase in student learning (Figure 1). Most improvement was achieved in understanding the role that DEQ plays in enforcing water quality standards (question # 3); the purpose of a standard (question # 7) and calibration curve (question # 10); what type of instrumentation is best suited to detect presence of zinc in water (question # 11); and the purpose of the "purge-and-trap" in VOC analysis.

As a further part of the assessment process, students compose a reflection paper of at least 500 words in which they address what they liked and disliked about their experience and what they learned not only about specific chemistry concepts or procedures but also about their ability to apply their knowledge to the solution of practical problems.

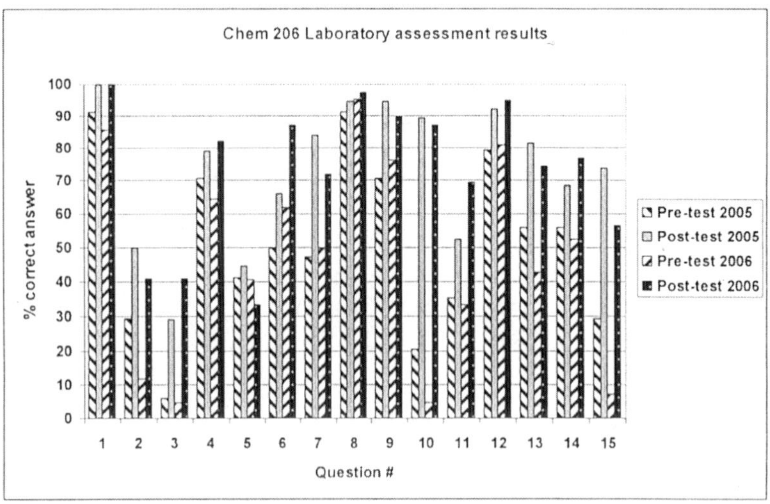

Figure 1. Comparison of pre-test and post-test assessment results

In analyzing students' reflections, some common themes seem to clearly emerge. In general students find the service-learning experience very enjoyable and valuable. They particularly like conducting sampling and testing in the field and having the opportunity to "get out of the laboratory." Connecting the analyses to a specific place seems to better define the analytical problem that they are posed to solve. One student wrote "Being able to go out to the site, knowing where the samples came from and see what the environment was really helped me understand the problem, what we were doing." They also feel this type of experience helps gain a deeper understanding of what they are learning and improve retention of their knowledge. By having a sense of purpose and knowing that the community partner will make use of the data, students feel much more accountable for their own work. One student wrote "I soon realized that the results of my analyses could make a big impact on future decisions. I was no longer doing simple experiments where the outcome is already known. I really liked this project because there was a purpose for what I was doing."

Students also commented that they gained a more realistic feeling of what an analytical or environmental chemist does. Many have no idea of the work involved in choosing an appropriate method of analysis or even gathering the reagents and equipment necessary to perform an experiment or calibrating instrumentation. Nor are they aware of the importance of keeping rigorous labnotebook records. Students develop a deeper awareness of the impact of human practices on the environment and the existence of regulatory policies that try to limit such impact. On the negative side, some students shared concerns about the complexity of analyses and the reliability of their results. Particularly upper-division students involved in environmental analyses for the CTUIR felt they did not have adequate time to produce reliable results and wished they had had the opportunity to repeat some of the tests. Others felt too much was expected given the timeframe of the course and that more background material should have been provided. For students in the general chemistry course use of advanced instrumentation such as the GC-MS or the AA seemed daunting. A few resented having to prepare a final presentation for the community partners.

Challenges and Additional Outcomes

There are a number of inherent challenges associated with the implementation of service-learning projects. The choice of the right community partner and the type of projects is paramount to the success of the program. Sometimes it is difficult to identify projects that can be completed in one quarter or semester; what seems easy at first glance may turn into a very challenging endeavor. As an example, several projects of environmental nature originally discussed with CTUIR scientists had to be abandoned because they were either too lengthy or required resources that were not available. One way to avoid "surprises" later in the term is to ensure that students have done their homework earlier on and that they receive constant feedback from the instructor. The choice of community partner is also very important. The partner has to understand the constraints placed by time limitations and students' skills on the significance of results. They also have to be willing to become an integral component of the service-learning experience and to be both learners and teachers. The partner is not a "client" receiving a "service" but rather an entity that actively contributes of the enrichment of the learning experience. Finally, class size has to be small to provide students with the adequate amount of assistance, particularly during the analysis phase.

One of the unexpected successes of introducing service-learning projects in analytical courses is the impact on recruitment and retention of students. Although the implementation of service-learning experiences in chemistry at EOU is relatively new and no long-term longitudinal studies have been completed, preliminary data suggest that it does have a positive impact on the number of declared majors. The number of declared majors increased by 28% after the first year of service-learning courses implementation (from 2002 to 2003) and has been steadily growing with an increase of 58% from 2004 to 2005 (after implementation of the service-learning project with GRMW in the third term of the general chemistry sequence). Similar results have been recently reported at Kalamazoo College (*10*) where the percentage of students declaring a chemistry major rose from 14% prior to the integration of the service-learning project to 26% and 21% after its implementation. Another measure of success is the increase in the number of students wanting to pursue independent research projects. As students are turned on to scientific investigation, they often express the desire to be involved in research whether connected or not to the original project

Acknowledgements

The author wishes to thank all EOU students who over the years have participated in service-learning projects, Dr. Ron Kelley and the community partners for their assistance and support. Particularly, I wish to thank Coby Menton and Jeff Oveson of GRMW and John Cox of CTUIR for their help in carrying out the projects. This paper was made possible through WRITE ON!, a writing retreat for Math and Science faculty sponsored by the Oregon Collaborative for Excellence in the Preparation of Teachers (OCEPT and OCEPT II), funded by a National Science Foundation grants DUE-9996453 and DUE-0222552.

The author wishes to thank Diane Smith and Elaine Jane Cole for their invaluable consultation during the writing of this manuscript. Support from the National Science Foundation (CCLI-0127120 and MRI-0116971) is gratefully acknowledged.

References

1. Claus,J; Ogden, C. In *Service-learning for Youth Empowerment and Social Change*; Claus, J. and Ogden, C. Eds.; Peter Lang, New York, NY, 1999, Vol. 5, pp. 1-6.
2. *Service-learning in Higher Education: Concepts and Practices*, Jacoby, B. and Associates, Eds.; Jossey-Bass: San Francisco, CA, 1996.
3. Hope, W.W.; Johnson, L.P. *Anal. Chem.* **2000**, *72*, 460A-467A.
4. American Chemical Society, Experiential Programs in Chemistry. http://www.chemistry.org/portal/a/c/s/1/acsdisplay.html?id=b6fc0848354d11d7e84a6ed9fe800100 (accessed December 7, 2006)
5. Wiegand, D.; Strait, M. *J. Chem. Educ.* **2000**, *77*, 1538-1539.
6. American Chemical Society, Committee on Professional Training. http://www.chemistry.org/portal/a/c/s/1/acsdisplay.html?DOC=education\cpt\Proposed_Revision_to_the_ACS_Guidelines.html (accessed December 7, 2006)
7. *Acting locally: Concepts and Models for Service-Learning in Environmental Studies*; Ward, H. Ed.; American Association for Higher Education: Washington, DC, 1999.
8. Stulas, A.A.;Clary, G.; Snyder, M. Service-Learning: Who benefits and Why, Social Policy Report: Society for Reserch in Child Development, **1999**, Vol. XIII (4), 1-20.
9. Astin, A.; Vogelgesang, L.J.; Ikeda, E.K; Yee, J.A. How Service-learning Affects Students; Higher Education Research Institute, University of California Los Angeles, CA January 2000; Executive Summary.
10. Esson, J.M.; Stevens-Truss, R.; Thomas, A. *J. Chem. Educ.* **2005**, *82*, 1168-1173.
11. Hatcher, M.E.; Aragon, E.P. *J. Chem. Educ.* **2002**, *79*, 462-464.
12. Kesner, L.; Eyring, E.M. *J. Chem. Educ.* **1999**, *76*, 920-923.
13. Fitch, A.; Wang,Y.; Mellican, S.; Macha, S. *Anal. Chem.* **1996**, *68*, 727A-731A.
14. Ram, P. *J. Chem. Educ.* **1999**, *76*, 1122-1126.

15. Hope, W.H; Johnson, L.P. *Anal. Chem.* **2000**, *72*, 460A-467A.
16. Werner, T.C.; Tobiessen, P.; Lou, K. *Anal. Chem.* **2001**, *73*, 84A-87A.
17. Tyson, J. *J. Math. Sci: Collab. Explorations* **2001**, *4*, 71-83.
18. Draper, A. *J. Chem. Educ.* **2004**, *81*, 221-224.
19. Gardella, J. *Community Access to Environmental Chemical Analysis.* http://www.acsu.buffalo.edu/~gardella/caai.htm (accessed December 7, 2006)
20. *The National Service-Learning Clearinghouse, Chemistry resource site.* http://search.servicelearning.org/index.php?q=chemistry (accessed December 7, 2006)
21. *Analytical Science Digital Librar.,* http://www.asdlib.org/ (accessed December 7, 2006)
22. *Monitoring and Assessing Water Quality, Volunteer Stream Monitoring: A Methods Manual.* http://www.epa.gov/owow/monitoring/volunteer/stream/ (accessed December 7, 2006)
23. APHA 1992. Standard methods for the examination of water and wastewater. 18th ed. American Public Health Association, Washington, DC.
24. *Handbook of Instrumental Techniques for Analytical Chemistry*, Settle, F., Ed.; Prentice Hall, 1997.
25. EPA Method 524.2 rev. 4.1, Purgeable Organic Compounds by Capillary Column GC/Mass Spectrometry; Methods for the Determination of Organic Compounds in Drinking Water-Supplement III (EPA/600/R-95-131)
26. USEPA. 1985. *Test methods for Escherichia coli and enterococci in water by the membrane filter procedure (Method #1103.1).* EPA 600/4-85-076. U.S. Environmental Protection Agency, Environmental Monitoring and Support Laboratory, Cincinnati, OH.
27. Harris, D. C. *Quantitative Chemical Analysis*, Sixth Edition, W.H. Freeman and Company, 2003.
28. EPA Method 7420, SW-846, 3rd ed., Nov 1986.

Chapter 10

Web-Based Animation in Analytical Chemistry

Thomas G. Chasteen

Department of Chemistry, Sam Houston State University, Huntsville, TX 77341

For the purpose of this chapter I will use active learning animations and interactive media to refer to content available on a computer that involves interactive learning by users as they view material above and beyond static text and images. Specifically this will refer to video or animations that change by way of programming or user input. For example, this could involve virtual laboratories—aimed at teaching about chemical reactions, animated instrumental schematics—aimed at teaching about the design and function of complex instrumental components and processes, or algorithm drilling—aimed at reinforcing computation steps such as determining a set of four quantum numbers for specific electronic configurations.

Why I use animations to teach analytical chemistry

In a very general way educators choose to go beyond textbook material because they feel that their students benefit from the alternate material they choose. Augmenting and improving the chances for students to learn is our job as educators and few instructors I know merely settle for *only* assigning expensive textbooks as the only source of course material.

Outside sources of material—beyond assigned textbooks—have, of course, always been used in analytical chemistry courses. Our analytical college texts have for generations incorporated outside readings from the scientific literature and schematics from instrument manufacturers. For instance, over fifty years ago, a text entitled *Instrumental Analysis* by Harley and Wiberley provided literature citations for up-to-date analytical methods in the literature (from *Analytical Chemistry*, *Analyst*, *Journal Optical Society of America*, for instance) and line drawings for "modern" instrumentation provided by their manufacturers (for instance, General Electric or Perkin-Elmer) (*1*).

All modern analytical chemistry texts incorporate these and more. CD-based animations have yet to take hold in analytical chemical texts like they have in freshman chemistry texts; however, partial color figures, full color plates, and instrument schematics are widespread.

The gap as far as I see it, lies between the information in a static image and that of a moving animation, a video, or an interactive program such as a Macromedia Flash-based program, for instance. The later items provide a much richer often more thorough and almost always more interesting source of the complex material in an analytical chemistry course.

A brief history of analytical-based animation

Molecular Modeling

Chemical animations visualized using mainframe computers (1964) were, obviously, the first available animations but were, equally obviously, not widely available to students. That history therefore and that of the later work-station-based software will not be included here, but is available (*2*).

Since all chemists, including analytical chemists, are introduced to molecular formulae, structure, and bonding early in their education, animated molecular visualization programs were some of the first to appear as personal computers became affordable and began to appear in school computer labs. A computer program called RasMol (**Ras**ter **Mol**ecules) was programmed at the outset for the UNIX platform and later the Windows and Macintosh operating systems. This molecular graphics software allows the user to view molecules on a computer screen in many user-selectable formats such as wire frame, ball and stick, space fill etc. A RasMol-generated ball and stick model of aspirin can be seen in the adjacent figure.

RasMol's author, Roger Sayle, initially at Imperial College, University of London and later University of Edinburgh, released a freeware version of the software in 1993, and put the code in the public domain. It has since been modified and updated, most notably to yield Chime. Publicly-available web sites with large RasMol-compatible molecular libraries are still available on the Web.

The public domain code of RasMol was adapted for Chime (Chemical mime) which was produced by programmers at MDL Information Systems in 1996 (*2*). Unlike the standalone program RasMol, Chime works as a plug-in in Internet browsers. The advantages of Chime over RasMol have been discussed (*3*) but briefly involve among others, higher-resolution molecular representation, better manipulation controls, and hardware independent movements. Open source stand-alone molecular viewers are also available (*4*).

The *Journal of Chemical Education*'s WebWare site has more complex versions of this type of software (all free) including a particularly versatile modeling program for molecular vibrations called 3D NormalModes Shockwave (*5*) that plays strictly in a Shockwave-enabled browser. Programs like these are supplanting QuickTime movie-linked IR spectra of common molecules available much earlier on the web (*6*).

Publisher-generated animations

As teachers and students became more familiar with these sorts of generalized animation experiences, textbook publishers began to include CD-based programs along with their support material for textbooks. One of the first useful analytical teaching animations that I ran into was an acid/base titration program initially written as a java applet. This interactive animation displayed a simple beaker and burette drawn on the screen, and the program allowed the user to pick from multiple base concentrations, different indicators, and different unknown acids. The user clicked on virtual buttons on the screen and added base until the endpoint was reached as represented by a change in the color of the solution being titrated. In a classroom setting this was a useful tool to complete many different titrations quickly for a large audience, requiring only a click of the reset button required to start the titration over. As an instructor I could "heat up" the discussion by proclaiming that a triprotic acid was being titrated by a dihydroxy base, quickly perform the titration, and ask for the students to calculate the concentration of the unknown acid.

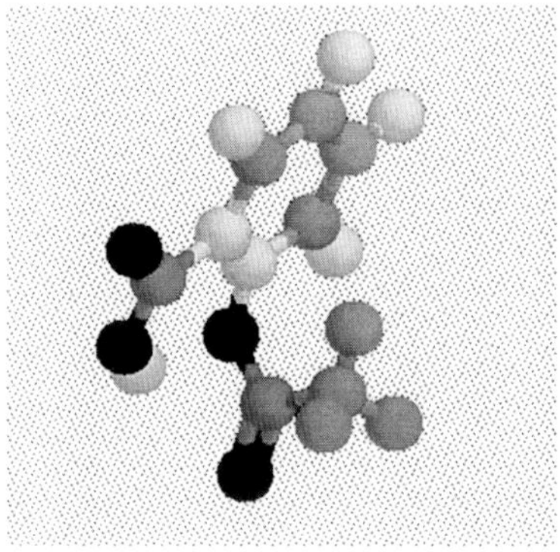

Figure 1. A RasMol-generated structure of aspirin.

Also available on publisher-generated CDs were, among many others, early kinds of (specifically) analytical-based animations involving Macromedia Flash-based or Java script-based "roll over" or "mouse over" functions: Static figures from the textbook were augmented with additional text that appeared when the user moved the cursor over a particular part of the image. While hardly exciting now, these initial efforts to provide computer-based content probably brought many computer-hesitant instructors into the fold of multimedia users in teaching chemistry and furthered the incorporation of computers as chemical teaching tools.

A worthy mention of Daniel Harris' Quantitative Chemical Analysis (Freeman Publishers) is appropriate here. His extensive inclusion of spreadsheet-based calculators for quantitative analysis courses was a very positive supplement to his text books in the 1990's; however, these were not animations and so are beyond the scope of this chapter. So-called living graphs —with interactive variable adjustments for determination of, for instance, the maximum wavelength for blackbody radiators or displayed Gaussian curves as well as many spreadsheet examples—are now maintained by his publisher on the textbook's web site (bcs.whfreeman.com/qca).

Video-based programs such as RealPlayer are also available and can use QuickTime media as source files; however, as of the summer of 2006, QuickTime files with Flash tracks can not be imported into RealPlayer files so content must be authored as a bit-mapped video track for use via RealPlayer.

A primer on animation construction programs

For some of us, who saw the complexity of material being failed by static textbook images, the next step was to create animated or at least augmented teaching content ourselves. For this chapter to be widely useful to readers—most of whom I assume are chemists and teachers—I will describe the computer programs required to produce content (I will call them authoring programs from here on) that are, in general the most easily obtained and most widely used. For instance, Macromedia sells software programs called Authorware and Director designed to author animated content. Avid Technology and Apple Corp. have multiple video editing suites. While these programs are very powerful, they are expensive and are also somewhat difficult to learn and therefore will not be viable authoring programs for most of the readers of this chapter.

Conversely Macromedia also sells a program called Flash (US$249) which is much simpler to use and which provides animation tools suitable to generate the content described below. This program is widely used for animated content. This therefore will lead me to describe the use of Flash software as an animation tool but not, for instance, Authorware.

Also not discussed here are audio-augmented animations. Both Windows Media, QuickTime, and Flash among others—and all video editing suites—allow audio to play along with animations. And with judicious compression of the audio content, relatively small file sizes can be achieved although audio greatly increases file size.

As animation programs evolve—compare those used in 1998 (*7*) to those of today, for instance—new programs will become available and old ones discontinued (*8, 9*). Therefore obviously not all programs used for creating animated content are covered or even be mentioned here. This chapter's purpose is to introduce the components of what we as teachers have experimented with to teach chemical/analytical concepts, and these ideas are what will last, not the programs that help us express them.

GIF Animators and QuickTime versus vector-based animation programs

But before I describe the use of vector-based programs, I would be remiss if I did not briefly describe bit-mapped animation programs usually called GIF animators (GIF stands for **G**raphic **I**nterchange **F**ormat). These generally very inexpensive stand-alone programs (freeware, US$25, etc.)—or the GIF animation export features in other programs—generate files that will play in any web browser as a series of individual images. The timing of these images is set when the file is generated; however, the bit-mapped nature of this format means that the content can contain **no** interactive elements (such as interactive button or internal links) and can not easily be resized for larger screens without degradation in image quality. These limitations are inherent in the bit-mapped format and are one of the reasons why vector-based (resizable) graphics are so much more widely applicable for animations not involving video. File size is another reason: the file size of a 30 second, vector-based animation of a moving object is normally smaller than a GIF animation of comparable image resolution and size, much smaller. GIF animation format also has a limited number of colors available.

QuickTime was a multimedia format created by Apple corporation initially for the Macintosh operating system 15 years ago, but which has been Windows-compatible for about 4 years. Early (>6 years ago) version of QuickTime (www.apple.com/quicktime) did not support the importation of Flash code (see below). This meant that to author teaching animations (that were not video) using QuickTime, individual frames had to be drawn and then imported into the QuickTime authoring program. And while internal links were possible (this, a form of navigation), their use was difficult, so the user was mostly limited to using the scrub bar to move around in the animation (see below). QuickTime 4 (~2000) allowed for the first time importation of a Flash track.

Important aspects of animations

Another powerful characteristic of vector-based animators—as opposed to bit-mapped GIF animations—is the use of different type faces and font size. Display/text fonts can be stored in the output file for use when the animation runs or can be set to display using the type face on the user's computer closest to the font authored in the file. This makes for clearer text and more consistent animations displayed across different computer platforms. And again, the resizing of vector objects (if the author allows an animation's window to be resized) will produce perfectly resized results because the objects in this format are scalable. With the wide range of computer monitors (and with video pod casting on the horizon) resizing animations is as important as ever.

After going beyond the limited colors available in GIF animations (usually 216 colors, a set called the web palette), more modern animation programs can handle a wide range of colors and grey scale, and these too lead to a truer representation of the images that the author originally created. The 216 color web palette was originally designed to accommodate colors that could be reproduced by 8 bit video ($2^8 = 256$). The last 40 "missing colors" (256-216) in the set were different on Macs and PCs and so excluding them meant that the 216 that could be used in that palette would be the same across those platforms (*10*).

The use of color in these teaching tools is meant to augment the images and text presented by drawing the users eyes to specific locations, influencing what is viewed first, or relate two different parts of the animation together.

The pace or speed at which teaching animations run has became more and more important as the speed of PC microprocessors has increased. This used to be a non-issue because animations could be viewed as fast as the animation could be rendered on the screen and most users still spent time waiting. Now that very fast video cards and microprocessors are common place, animation authors must choose the playback speed of an animated sequence. In Macromedia Flash this is called the frame rate and controls the playback speed in reference to actual time instead of the clock speed of the microprocessor.

As I have advanced (if I might use that word) in the analytical teaching animations I have authored over the past 10 years, I have modified the characteristics of my animations in various ways (see below); however, one of the most significant changes I have made has been the incorporation of user-accessible navigation tools inside the animation. Though the scrub bar in the Windows Media Player and QuickTime players allows gross movement through an animation, the use of navigation buttons (easily programmable in Flash) are a more accurate means of navigation. The adjacent figure of a QuickTime-based animation indicates the player's scrub bar. Navigation inside the animation using the player's controls is accomplished by clicking on and dragging the scrub bar. This software also allows the user to stop, restart and run the animation in reverse (this last only works if no stops are programmed by the author). The default setting for newly opened media players of this sort is to play through the entire animation continuously. This "passive user mode" displays the animation like a movie, and as I note below programming can be used to require the user to interact with the content in hopes of increased user learning.

The animation's navigation buttons can also be seen on the right side in that figure. Other types of useful interactive navigation tools are hypertext links inside animation text that lead the user to other parts of the animation. Using specific colors for this text helps the users to know what to click.

User interaction with navigation buttons is just one way to pull users into the content. Another method that I have used is to short circuit the passive user mode by providing frames of content that display for times that are too short to read all the content. This forces users to invoke the stop and resume buttons (as well as navigation buttons, see adjacent figure) to be able to display the content that they want to access (*11*).

Copyright Issues

The issue of copyright comes mainly to those who author their animations and place them on the web. Web-based animations are copyrighted from the time they are posted by the author, even though they have not been registered with the U.S. copyright office (*12, 13*). With that said it has been this author's experience that web-available animations posted on my local server have been widely used (my animation directory gets about 170,000 hits per year), and I routinely

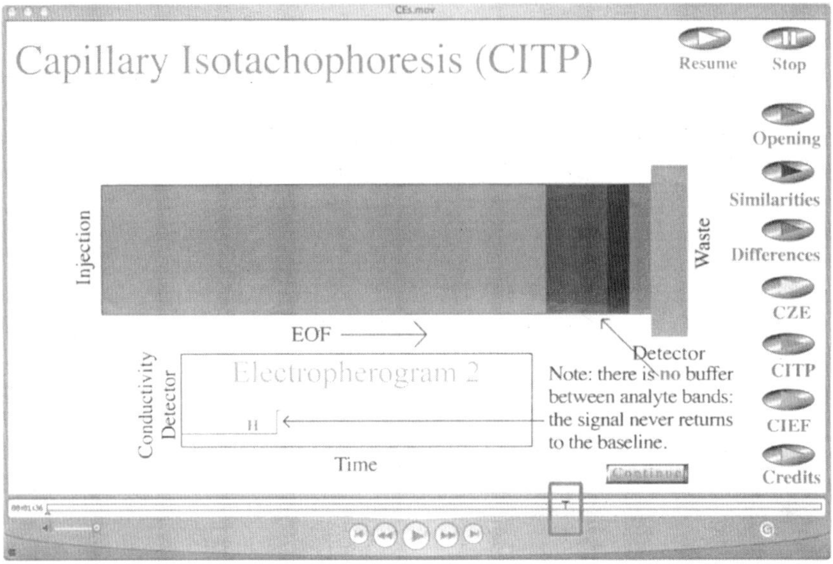

Figure 2. A screen shot of a QuickTime animation with parts of the scrub bar boxed in red.

receive copyright release requests—mostly from academics—to incorporate my animations in on-line presentations or talks. My policy has always been that users can download my animations to their personal computers (for use in talks and PowerPoint presentations) but I pointedly ask those who request it to *not* upload my animations to their publicly-available servers. In this way I can keep some control over my material. And a quick Google search for one of my most popular analytical animations (over GC/MS) shows that the file has not been uploaded to any other "Google-findable" server—unless, of course, my file's name has been changed.

Examples of Analytical Animations

Static/passive use animations as teaching tools

A short word here about the kinds of animations I will focus on below. Instead of molecular modeling which involve so-called visospatial images (images like 3-D molecules or maps) most of what I will discuss is not inherently visual and instead describes relationships, interactions and schema (*14*). It was these kinds of concepts in analytical chemistry that I have been struggling to teach.

The very nature of this topic, involving the word *interactive*, leads one toward a teaching tool beyond simple pictures. Remember that I and at least a few others (*7, 11, 15*) believe that the failure of our chemistry text's static images has led many of us to author animations in the first place (see above). Therefore little time will be spent discussing static tools. It should be noted that color images in texts have gotten much better in the past 10 years as opposed to even all the previous 25 years before that. It is easy to argue that this improvement is powered by the increased book costs and the growing interest in publisher profits from college textbooks. And, as noted above, publishers are now routinely including textbook content on free CDs given to textbook adopters. So instructors can include images in PowerPoint talks either directly supplied by the publisher or with image imported from the publishers' discs.

Teaching animations created before vector-based programs became widely used simply involved bit-mapped images as a sequence of frames. A series of static images, individually drawn and saved, were strung along sequentially as an animation. Programs such a QuickTime, Animation Pro (bundled with PaintShop Pro) or Video for Windows (later Active Movie, later DirectShow, now Windows Media Player) were often used to size and time the frames of the animation and the software player's scrub bar was used to simply navigate. As noted above, this was how many early (nonprofessional) teaching animations were constructed and played back; however, this was at least one step better than GIF animations because the playback timing and frame size and color set could be more easily controlled, and since the format in which the files were saved dictated the programs that were going to play them (.mov, .wmv, .avi, .asf, etc.), the old GIF animation's requirement that the file had to play in a browser window did not apply. The advent of Macromedia Flash changed all this because of the cross platform applicability and ease of the incorporation of buttons and hypertext links. But with that said, passive movie animations as teaching tools can still be a vast improvement over GIF animations because they can be stopped, started, and easily rewound and—depending on the software used to create them—can display almost any level of complexity an author wants to use. Dr. Marian Hyman at Texas A&M University has a dozen passive instrumental analysis movies (in Window Media Video format) available on-line in a public directory (*16*).

Interactive animations

In its simplest form, an interactive animation may not incorporate any movement at all, but instead involves an interaction with the software by the user based on what's displayed. Previous research has suggested that this clearly benefits learners (*14, 17*). David Brooks at University of Nebraska, Lincoln has recently described an on-going automated practice system that uses interactive image maps that work this way (*18*). Questions posed on the screen require a response from the user, such as clicking on a specific region of a graph, data set, or schematic. This action elicits feedback to the screen. Simple but potentially powerful. For instance, mass spectral data could be handled in this way to prompt the user to click on the molecular ion or *m/z* peaks for specific fragments with appropriate feedback for correct or incorrect selections.

The next level of interaction can be a more hands-on approach such as that used by Conrad Trumbore and coworkers at University of Delaware. The adjacent figure is a screenshot displaying an Flash-based animation that depicts the interaction of infrared light with various atmospheric gases (*19*).

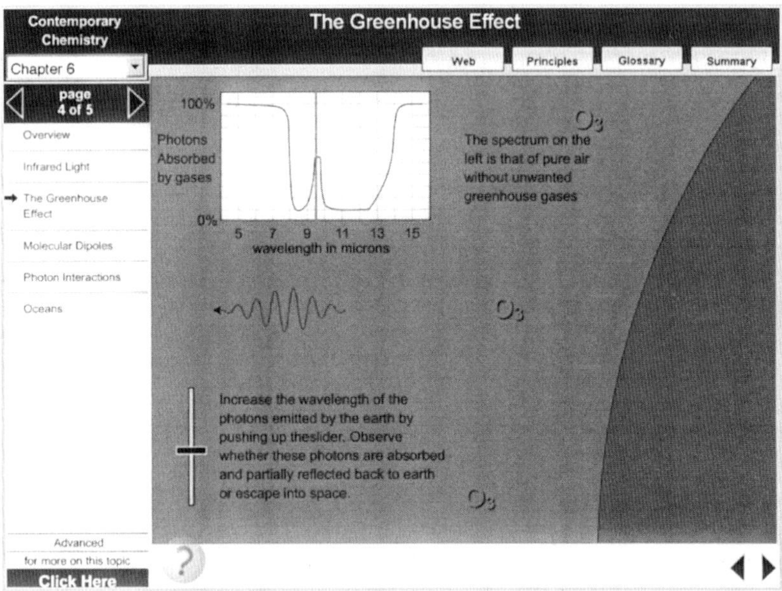

Figure 3. A screen shot of a Flash-based interactive animation depicting the interaction of outgoing IR radiation with atmospheric gases.

Since the analytical environmental chemists wants to know which IR wavelengths interact with which gases, the authors have supplied a user-controllable slider which "scans through" an IR spectrum while the animation displays both an image of an IR photon and a cartoon of the interaction of the appropriate gas floating in the earth's atmosphere. This allows the users to connect specific parts of the atmosphere's thermal IR region with the atmospheric components that absorb outgoing IR at those wavelengths.

An even more hands-on animation (and one decidedly more analytical than the last example) has been authored by Thomas Greenbowe at Iowa State (*20*). This is a Flash-based virtual titration, and the animation is complete with a solution-delivering burette, three different redox reaction choices, and even a spinning stir bar and color changes in the titrating flask. This problem does, in an interesting and engaging manner, just what we as teachers have asked our student to do to master the concept of redox reactions: perform multiple different redox reactions with different reagents and do it multiple times. And while this specific interactive animation doesn't require that students balance the reaction before starting, it could easily be engineered to do so or could also be programmed to not display the correctly balanced reaction unless the student asks for a hint. Using an external database, the software could also keep track of which students ask for which hits, etc.

In the future many of our chemistry laboratories may be influenced by the use of virtual labs of this sort, and, in my mind, to the advantage of our students. Commercial programmers are, of course, programming software products that do what Greenbowe has done (above) and more. Late Nite Labs has an entire laboratory suite involving enough labs for much more than one undergraduate laboratory semester (*21*) that can be maintained on the company's server and to which student access is sold for their use by teachers. Though this is one of the few commercial products I will mention in this chapter, there are many others available now and more coming. For instance, another is Virtual ChemLab (vig.prenhall.com). Figure 4 displays many of the characteristics of a virtual laboratory of this complexity.

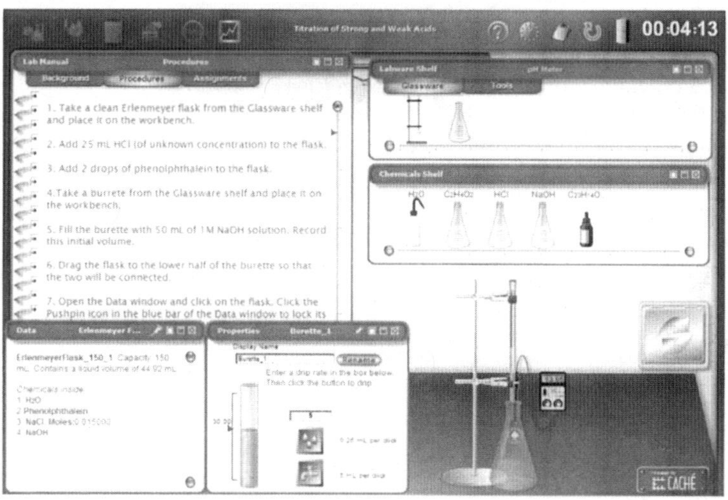

Figure 4. A screen shot of a Flash-based interactive animation of a virtual lab bench with laboratory shelf, chemical shelf, lab manual, and data record.

Virtual glassware such as beakers, flasks, and graduated cylinders are presented on a virtual laboratory shelf from which they can be dragged onto the virtual lab bench for use in different experiments. Virtual chemicals are available on a virtual chemical shelf. Programmed feedbacks that the users get using the software include clicks for thermometers "attached" to beakers, color changes in chemical precipitates in test tubes, reminders of maximum container volumes if the analyst tries to over fill them, etc. Data recording functions such as temperature, amounts of reagents added, pressure in closed containers, etc. are available depending on the lab under study. Data manipulation such as data plotting (such as temperature, titration volumes, pH, solubility curves, etc.) are also accessible. The Flash-based format of the software means that students need only use an up-to-date web browser.

In my college senior level instrumental course (which I have been teaching at SHSU for 15 years) I have struggled in the past to teach some of the more complex instrumental methods of analysis. The major textbook in this field (Skoog, Holler, and Neiman's *Principles of Instrumental Analysis*) is an excellent resource though its last edition (published in 1998) is getting a bit long in the tooth. (The next edition is due in late 2006). With that said, again, it was what I viewed as the failure of the text's static schematics and figures that led me, at first, to author simple animations that displayed light paths, gas flows, and data display (that is, not visuospatial relationships). As I gained in (simple) programming knowledge my initially bit-mapped QuickTime movies grew into Flash animations. See Figure 5.

The immediate benefit of this was resizability of the animations displays and a steep reduction in file size. As I learned how my students interacted with my animations—based on feedback from them in my classes—I improved the navigational tools and also stopped making animations that played continuously from end to end, the passive, non-interactive route.

Instead I authored animations that automatically stopped on a specific frame and required the user to interact with the animations to proceed, go backwards, go forwards or go to another section of the animation all together. The adjacent figure is a screen shot from an animation involving gas chromatography/mass spectrometry (GC/MS). The animation covers all the instrumental parts and functions of this widely used analytical instrument and requires the user to advance though or navigate around in the animations "scenes" using the internal navigation buttons, scrub bar, or keyboard. The animation contains over 5000 frames and 11 scenes.

Assessment

Like all of the chemistry animations authors I have contacted as I worked on this chapter, hardly any of us have performed a statistical evaluation on the

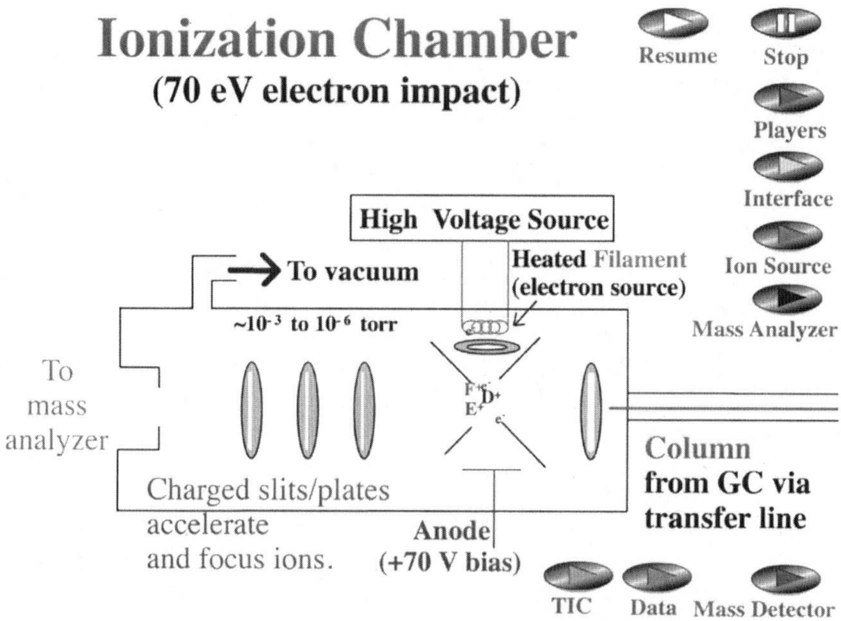

Figure 5. A screen shot of a Flash-based interactive animation on gas chromatography/mass spectrometry.

success of our animations in teaching what we have produced the animation to teach (see below). I have, however, performed an experiment using an interactive Flash animation that I have authored to teach the concept of quantum numbers to students (yes this is not really analytical chemistry either). This involved one 50-minute lecture in quantum number generation, taught after electron configuration in a freshman, majors chemistry class. Two class sections in the same semester (42 and 45 students) were evaluated and although both sections were given the same lecture material on quantum numbers (and an on-line quiz over that material), one section (n=45) was assigned the quantum number animation. The other class section (n=42) also saw the single class lecture but was not assigned or even told of the existence of the animation. Those assigned the animation could not access the required quiz until they had accessed the animation. Both sections were given the identical questions on quantum numbers on the next in-class test and their scores on those test questions tabulated. The differences in tabulated scores between these class sections over just the quantum number material were statistically insignificant, at 95% confidence level (*22*). The differences in their on-line quantum number quiz scores were, by the way, also statistically insignificant.

A recent review by Tversky et al. of pedagogical studies using animations as teaching tools—many in science and math—concluded that the purported advantage of animation over static images apparently stems in large part, if not completely, from the fact that animation authors routinely add additional information or steps in the ideas conveyed or include interactivity. They decided that if the static images to which the animations were compared contained the same content, the static image would teach with equal effectiveness (*14*).

Brian Pankuch at Union County College (Cranford, NJ) noted in a discussion with me this spring that he found students who did well on computer-based programs (that he has been using for over 10 years) *did* perform statistically significantly better on multiple choice questions over material covered by those programs. And this was true even if they did poorly on material on the same test *not* covered by the interactive programs. But he noted that the program users also were probably spending more time studying the material than students not using the computer-based programs. His conclusion was "that spending additional time on problem solving pays off in substantially better grades. But that time spent on any type of problem solving is somewhat effective. So I suggest that using a myriad of methods to get students practicing and thinking about the material is effective, and once the novelty of a given method wears off its effectiveness is less. So it is good not to spend too much time trying to perfect a given method—animation for instance—better to keep making more methods available."

I find Dr. Pankuch's observations a thoughtful and practical view of the way that I also think students learn. But I would also note that I concur with *both* of the above positions, his and that of Tversky et al.'s review: 1) Using animations in teaching provides an advantage to students who interact with the material through increased study time and focus; and 2) authoring animations provides the instructor another chance to focus, to think about the material more in depth, and to provide more content for learning, and this even though the result may not necessarily teach students better than static materials if we could just get students to dedicate the time to engage with that material. But from the point of view of engaging students in hopes of increasing their time focusing on the material, animations may do this by piquing the interest of students bored with static tools (*23*). And finally, the variations in the way different students learn the same material (*17*) is addressed by having different media covering the same content.

The Future of Animations in Analytical Chemistry

As someone who has been watching the development of computer-based animations for a generation and authoring them for over a decade, it is clear to me where the future in this area of teaching is going. Just as we, as teachers, have to a large degree adopted publisher-based text books, and just as we, to a smaller degree, have adopted publisher-based overhead transparencies or PowerPoint presentations, and finally, just as we have to varying degrees adopted publisher-supplied test banks for our online course management software, the future for computer-based animations lies with the publishers of our academic content. To a lesser degree, free-standing commercial products will also find more of a place in our teaching arsenal. More and more of our text books will offer more complex and more interactive animations on their support CDs. And more and more authors will use these materials to teach. As I noted above, one of the prominent instrumental analysis books in the field of analytical chemistry will be issued late in 2006. For the first time it has a companion CD and website....

Along with publisher-generated companion CDs containing high quality animations, commercial vendors also offer CD- or video-based teaching materials that incorporate animations; however, analytical content is usually not a major portion and upper division material is scarce. One of these is Academy Savant (www.academysavant.com). While the cost of these materials purchased individually is steep (compared to, for instance, textbook prices), incorporating site licenses and/or academic pricing will probably make these materials more affordable in the future. And although I have not seen this yet, I am sure publishers have struck deals and are already bundling licenses to commercial content sites (that is, not publisher-generated sites) with their textbooks.

Acknowledgements

Computer support of the animations authored and described here provided by Dr. James Van Roekel in SHSU's Office of Academic Instructional Technology and Distance Learning. T.G.C. gratefully acknowledges the support of the Robert A. Welch Foundation and the Sam Houston State University's Research Enhancement Fund.

References

1. Harley, J. H.; Wiberley, S. E. *Instrumental Analysis*; Wiley and Sons: New York, NY, 1954.
2. Martz, E.; Francoeur, E. *History of Visualization of Biological Macromolecules*, URL http://www.umass.edu/microbio/rasmol/history.htm.

3. Martz, E. Chime vs. RasMol, http://www.umass.edu/microbio/rasmol/history.htm (accessed Dec 8, 2006).
4. JMol. http://jmol.sourceforge.net (accessed Dec 8, 2006).
5. Charistos, N.D. Tsipis, C.A.; Sigalas, M. P. 3D Normal Modes Shockwave. http://www.jce.divched.org/JCEDLib/WebWare/collection/reviewed/WW022/index.html (accessed Dec 8, 2006).
6. Ellsworth J.; Stramecki, B.; Parrill, A. Animated Infra Red Spectroscopy. http://www.cem.msu.edu/~parrill/AIRS/functional_group.html (accessed Dec 8, 2006).
7. Burke, K. A.; Greenbowe, T. J.; Windschitl, M. A. *J. Chem. Educ.* **1998**, *75*, 1658.
8. CHEMINFO Indiana University. Molecular Visualization Tools & Site. http://www.indiana.edu/~cheminfo/ca_mvts.html (accessed Dec 8, 2006).
9. Ojamäe, L.; Hermansson, K. MovieMol. http://www.chemistry.ohio-state.edu/~lars/moviemol.html (accessed Dec 8, 2006).
10. Weinman, L. The browser-safe web palette. http://www.lynda.com/hex.asp (accessed Dec 8, 2006).
11. Chasteen, T.G. *J. Chem. Educ.* **2001**, *78*, 1144.
12. U.S. Copyright Office. Copyright Registration for Online Works. http://www.copyright.gov/circs/circ66.pdf (accessed Dec 8, 2006).
13. U.S. Copyright Act. http://www.law.cornell.edu/wex/index.php/Copyright (accessed Dec 8, 2006).
14. Tversky, B; Morrison, J. B.; Betrancourt, M. *Int. J. Human-Computer Studies* **2002**, *57*, 247-262.
15. Zare, Richard N. *J. Chem. Educ.* **2002** *79,* 1290.
16. Hyman, M. Movie Directory. http://dist.stat.tamu.edu/pub/chem (accessed Dec 8, 2006).
17. Ferguson, E.L.; Hegarty, M. *Cognition Instruct.* **1995**, *13*, 129-160.
18. Brooks, D. The Evolution of an Automated Practice System. http://dwbrr.unl.edu/Demo/CourseDemo/startdemo.html (accessed Dec 8, 2006).
19. Trumbore, C. Using Computers in Chemical Education Using Computer in Chemical Education Newsletter, Fall 2005. http://home.eclipse.net/%7Epankuch/Newsletter/Pages_NewsF05/F2005_News.html (accessed Dec 8, 2006).
20. Greenbowe, T. Redox Titration in Acidic Medium. http://www.chem.iastate.edu/group/Greenbowe/sections/projectfolder/flashfiles/redoxNew/redox.swf (accessed Dec 8, 2006).

21. Late Nite Labs. http://latenitelabs.com (accessed Dec 8, 2006).
22. Harris, D. "Quantitative Chemical Analysis"; W. H. Freeman: New York, 2003; p. 70.
23. Charistos, N.D.; Teberekidis, V.I; Tsipis, C. A.; Sigalas, M.P. *Educ. Inf. Technol.* **2003**, *8*, 369-379.

Chapter 11

Progression of Chemometrics in Research Supportive Curricula: Preparing for the Demands of Society

John H. Kalivas

Department of Chemistry, Idaho State University, Pocatello, ID 83209

This chapter describes teaching approaches used in the three laboratory courses general chemistry, quantitative analysis, and instrumental analysis that assist students to attain expertise with process skills needed for success. The courses use active learning mechanisms to develop research skills in concert with requirements common to research laboratories. The progression of chemometric tools used is traced through the three courses. From the courses, students gain experience in thinking scientifically with improved reasoning ability to make their own decisions with increased confidence, solving problems in teams by sharing knowledge, written and oral communication of scientific concepts, and information gathering techniques. Budgetary matters are also exemplified.

Introduction

Society, technology, and the world economy are becoming much more dynamic and changing at increasingly faster rates. Students need to be provided with opportunities to acquire not only knowledge, but process skills needed to survive and be successful in this environment (*1-10*). Traditional forms of lecture and laboratory courses often satisfy the factual knowledge needs of society, but usually lack in teaching students process skills such as critical thinking and problem solving. Other process skills needed are teamwork, communication, computer literacy, management, and assessment. Essentially, there are two parts to education: content and process and the process part is often not given sufficient consideration (*10-12*). By putting more emphasis on developing process skills through the use of active learning, students can become better learners, thinkers, and problem solvers and improve their success in the real world (*13,14*). Additionally, retention should be facilitated by engaging students with an active discovery mode of teaching.

Traditional laboratory courses are passive and are based on a laboratory manual with organized data tables and detailed procedural steps including directions regarding what data to obtain, when to make particular measurements, and step-by-step calculations. Little of this helps students learn the process of science and be able to deal with new situations or problems in the real-world.

For students to learn the process of science, a research supportive pedagogy is essential (*15-17*). The typical primary source of a research experience is a capstone undergraduate research (independent problems) course. Instead, research supportive curricula developing research (science) processes should be applicable throughout an undergraduate's education. Research skills needed include searching, reading, and evaluating literature, articulating a research question, designing an experiment, formulating scientific explanations using experimental evidence, and effectively communicating the results of a scientific investigation. These skills are also sought by industry, but are often found lacking (*3-8*). Thus, it appears that the capstone experience is not enough.

Interest is growing in developing methods to emphasize science processes in analytical chemistry. Developments include student centered active learning (*14,18*), open ended experiments (*19*), role-playing (*20*), and group projects with real world samples (*21-27*). This is typically accomplished by using problem based learning (PBL) and cooperative learning (CL) techniques in order to engage students in active learning during experiments (*28-32*). Essentially, little information is provided about how to perform experiments and students are expected to make group (team) decisions.

While awareness of chemometrics is not key to success in society, it is a knowledge area important to success in analytical chemistry and society greatly benefits from its use. Approaches to lecture courses on chemometrics have been described (*33-35*) including a tutorial (*36*) and use of MatLab (*37*). A chemometric module for an instrumental analysis course has been developed (*38*). Single experiments using chemometric tools have been made available and include using near infrared (NIR) spectroscopy with principal component regression (PCR) for fat and moisture content in live fish (*27*), multivariate spectrophotometric analysis of various systems with least squares (*39-41*), PCR

(*42*), and factor analysis (*43-45*), principal component analysis with IR data (*46*), atomic absorption spectrometry (*47*), and lanthanides (*48*), optimization by simulated annealing (49) and simplex (50-*57*), experimental design (*58-63*), sampling protocols (*64*), quality control/quality assurance (QC/QA) (65-*67*), and error analysis (*68*). This chapter traces the progression of incorporating chemometrics with PBL and CL methods to enhance research supportive curricula in general chemistry and quantitative and instrumental analysis laboratories.

General Chemistry

A primary chemometric tool that can be learned in general chemistry is univariate calibration with error analysis. Univariate calibration is fundamental not only to analytical chemistry, but all areas of chemistry and is easily incorporated into general chemistry.

Rather than drawing on current applications of chemistry to real-world situations, the two semester-long general chemistry laboratory courses at Idaho State University (ISU) have been restructured to utilize a guided inquiry approach (*69-73*). In the new structure, students work in teams to design appropriate experiments to answer a research or focus question and the lab emphasis is on the investigation process. The instructor uses guiding questions in the pre-lab phase of the laboratory session, to guide the class as a whole to develop an experimental design. Sometimes, teams generate team specific experimental designs. The next phase is the experimental component where the students perform the experiment. The last portion of the laboratory period is devoted to post-lab student presentations, discussions, and reflection with the possibility of a new experiment. Throughout the laboratory session, the instructor's primary role in the student-centered learning process is that of a facilitator, i.e., using questions to guide students to the answers. Some labs are case studies, which provide context rich problems in real-world settings (*74-78*). Such problems with more than one feature encourage thought, requiring students to identify and separate the features and organize essential information. Case studies often necessitate searching, reading, and evaluating literature, important parts of a research supportive curriculum.

In addition to the guided-inquiry laboratory experiments, each semester has a project component where in the first-semester teams develop lab proposals (with budgets) that describe guided inquiry labs that could be used in the course. Eventually, one of labs is actually used in the course. Service learning is a fairly new pedagogical component to chemical education (*79-84*). The intent is to provide the students with an opportunity to form connections between lecture/laboratory knowledge acquired by the students and an application of that knowledge to benefit the community. Sometimes this occurs simultaneously with a research experience. In the second semester, students work in their teams to develop and communicate an interactive concept discovery session to a kindergarten through sixth grade (K-6) class. By carrying out this service learning project, students improve their processing skills as they must search, read, and evaluate literature to articulate a research question and design an experiment to complete in a K-6 class. In the process, general chemistry students need to effectively communicate the results of a scientific investigation by formulating scientific explanations using the generated experimental evidence. The mechanics of the course in specific details are described in ref. (*85*).

Course Overview

During each semester of general chemistry laboratory at ISU, laboratory sections meet weekly for three hours with 20 students per section. All work is performed in teams of three to four students with maximum heterogeneity, e.g., gender, ethnicity, and scholastic abilities. In the first semester course, there are eight discovery oriented labs, one case study, the lab project, and the resultant student created discovery lab. Six weeks are allocated for team development and presentation of the lab proposal. During the second semester course, there are six discovery labs, four case study labs, and the service learning project. Six weeks are allocated for development and presentation of the service learning project. Grading protocol and student expectations are described in ref. (*85*).

Chemometrics

Some of the discovery labs require students to develop univariate linear models using least squares. For most students, this is their first introduction to modeling and the resultant error analysis of the model. For example, in the second lab of the first semester, the focus question is, "What is the mathematical relationship between Celsius and Fahrenheit as experimentally determined in lab?" From the lab they learn that plotting the two units allows the mathematical relationship (model) to be determined. In the second semester a lab has the research question, "What is the relationship between color, wavelength, absorbance, and concentration?" In essence, students discover Beer's law, a fundamental analytical chemistry relationship. While it is only univariate, they are now knowledgeable for chemometric extension to the multivariate case in quantitative and instrumental analysis.

Assessment

The described approach has not been formally assessed with tools such as SALG (student assessment of learning gains) (*86*). However, the teaching methodologies implemented are well documented (*87-90*) and have been used elsewhere with improved assessment results (*10-12,71,72,91,92*). Additionally, this teaching format meets the goals set out in various reports on curriculum issues *(1,2,93,94)*.

Quantitative Analysis

Laboratory courses offering classical methods of chemical analysis need to be maintained and taught at the university level (*6-8,23*). With the advent of instrumental analysis methods, there has been a gradual shift away from teaching classical methods in quantitative analysis. However, classical methods are needed in order to establish standards to calibrate instruments. For example, analysis of protein and fat content of food products are required for food labeling regulations. The Kjeldahl method, involving a titration (*95*) and solvent extraction coupled with gravimetry (*95*) are typically used to develop reference standards for protein and fat analysis, respectively. These reference standards can then be used to build spectroscopic multivariate calibration models with chemometrics. Thus, it is important for a modern quantitative analysis laboratory course to contain classical gravimetric and volumetric analysis exercises implemented with standard unknowns in order to gain expertise for potential uses as reference methods in chemometric multivariate calibration problems. Additionally, by carrying out classical unknown analyses, students are allowed to work independently and build confidence in their ability to obtain accurate and precise results. However, this traditional approach does not furnish a climate in which to build inquiry, develop problem solving skills as a team, cultivate leadership responsibilities, or improve both oral and written communication skills, all of which can be obtained by exploiting real-world analyses in PBL and CL formats.

The real-world approach has been demonstrated as an effective way to teach quantitative analysis (*18,20,26,96-98*). One strategy shown to be successful uses an ecosystem provided by an aquarium (*26,96,98,99*). In ref. (*96*), a tropical freshwater aquarium was used and potentiometric analyses were implemented. A tropical marine aquarium was used to teach gravimetry, titrimetry, electrometry, and spectrophotometry in reference (*26*). Rather than setting up an aquarium in the laboratory, a strontium analysis by atomic absorption has been detailed where students obtain marine aquarium samples from local pet shops

(*98*). However, a course designed on only a real-world basis lacks analysis of traditional unknowns commented on earlier. Most importantly, real-world samples are true unknowns and it is difficult to directly grade students on accuracy. Thus, essential laboratory skills as well as development of confidence in a persons ability to conduct a real-world analysis are probably not firmly established.

A course has been designed at ISU to incorporate ideas using both the traditional approach and the real-world approach. In this way, the two approaches are complementary thereby eliminating respective deficiencies. More information on the course design is available in ref. (*27*).

Course Overview

The sophomore laboratory sections meet twice weekly for three hours each with 16 students per section. The first ten weeks are devoted to individually performed traditional unknown analyses based on instructor-supplied unknowns. During the last six weeks, students carry out an ecosystem study of a cold water trout environment simulated by a trout aquarium. This work is performed in teams of four students with maximum heterogeneity. New teaching tools in the second session include costs analysis, QC/QA, and NIR spectroscopic analysis of live trout using multivariate calibration by PCR.

Classical Labs

Classical labs include a gravimetric determination of sulfate and lipid concentration in certified pork meat requiring a solvent extraction, EDTA titrations for calcium, followed by a student-designed calcium and magnesium EDTA analysis, and spectrophotometry for copper. In the self-designed experiment, students are essentially presented a sample containing calcium and magnesium and are told to report concentration values.

Ecosystem Study

Many physical and chemical factors affect trout, thereby providing an abundance of studies for students to conduct (*100*). Analytes include dissolved oxygen, alkalinity, nitrogen cycle components ammonia, nitrite, and nitrate, and lipid and moisture content in live trout muscle by NIR and chemometrics. To perform these analyses, team members rotate through a manager role, who is responsible for organization of a particular analysis, i.e., all students in a team manage at least one analysis. The laboratory instructor must be provided with an outline of individual task assignments before managers can begin directing analyses. Additionally, managers must also prepare a proposed budget. To assist students in preparing the budget, managers and technician wages with fringe benefits are supplied as well as costs for waste disposal and any specialty items not included in a Fisher Scientific catalog. While implementing analyses, managers keep track of actual costs and prepare final budgets. When all teams have completed ecosystem studies, team results are compared and graphed by the managers of respective analyses. With QC/QA, students ascertain the creditability of their analyses. Oral and written reports on the ecosystem are required from each team. To assist students in preparing reports, required discussion topics are furnished in the laboratory manual. As with general chemistry, the instructor acts as a facilitator, albeit in the quantitative analysis laboratory, the laboratory procedures are not in a guided inquiry format.

The trout aquarium is a dynamic system with the changing chemical balance. For example, by changing water temperature, students observe the direct influence on dissolved oxygen and experience science as researchers do. Working with the trout environment provides students the chance to collect data and formulate conclusions about the general health of the ecosystem which can then be extended to the health of wild trout populations. Students see the direct influence of natural variables on the ecosystem and gain an appreciation of chemistry's role in the environment and biological processes.

Chemometrics

The quantitative analysis course is often the first introduction to basic statistical concepts such as the Gaussian distribution, mean value, standard deviation, probability, confidence interval, control charts, propagation of error (error analysis), statistical testing of measured values, and least squares for univariate calibration (*101*). While this is not what is normally thought of as chemometrics, the concepts do represent needed material for understanding more complicated chemometric tools. Most importantly, once students understand Beer's law with univariate calibration, it is straightforward to make the jump to multivariate calibration with Beer's law. From this, students can grasp the fundamentals of rewriting Beer's law into the inverse form for multivariate calibration using PCR. Will the sophomore students completely understand PCR? Probably not. Instead, students learn to use PCR as tool to solve a complex problem. When students are in an employment situation with a similar calibration problem, it is hoped that PCR will be remembered as a solution.

Using PCR for the trout analysis allows students an opportunity to realize that complex sample matrices can be analyzed with the use of spectroscopy. It is also revealed that more than one analyte concentration can be predicted from the same spectrum. A principal component plot is utilized to show students where their trout sample lies with respect to calibration concentrations used to create the model. Through examination of this plot students determine if their result has been obtained by interpolation or extrapolation.

Traditional approaches for determining lipid and moisture require the death of the fish and for lipid content, hazardous waste is generated. In the course at ISU, trout are taken from the aquarium, analyzed for lipid and moisture contents using NIR spectroscopy with a previously built PCR model, and then returned to the aquarium. In this way, lipid and moisture determinations are accomplished in a manner of minutes and trout do not have to be destroyed.

Assessment

In the early stages of development, the course was assessed using a survey format developed specifically for the course by the Office of Institutional Research at ISU. The approach is similar to that used in SALG (*86*). Three assessments were performed; the first day of class, after the unknown analysis portion, and at the end of the class. Statistical analysis of the results showed that some process skills were gained by the traditional teaching approach and others were gained by the ecosystem portion. Key processes learned included critical thinking, finding solutions to new problems, working as a team, leading a team, communicating scientific concepts, and integrating knowledge across disciplines. Further details of the assessment results are in ref. (*27*).

Instrumental Analysis

A common approach to instrumental analysis laboratory is requiring students to use an instrument to perform an analysis of a provided single analyte unknown sample with no real-world connection. This passive teaching format is applied to as many instruments as possible during a term. Active learning is not promoted with this approach and hence, suffers from problems already discussed in this chapter. Rather than trying to expose students to as many instruments as possible, a different mind-set is to have students gain detailed knowledge and experience with a few instruments, i.e., depth of knowledge rather than breadth. While fewer experiments are performed, the emphasis is on the depth of the problem solving rather than the extensiveness of analytical methods. With an emphais on only a few instruments, it is easier to implement chemometrics.

Incorporating PBL and CL into instrumental analysis laboratory with real-world samples overcomes limitations with the traditional approach and allows students to undertake the complete process of performing an anlysis. For example, in a team-work environment, students identify a problem, collect samples, perform the sample workup and pretreatment, make measurements and calculations, validate results through QC/QA, and effectively communicate recommendations based on obtained results. This section describes a new approach to teaching instrumental analysis laboratory at ISU by providing students with a fundamental analytical curriculum. More information on the course design is available in ref. (*25*).

Course Overview

Instrumental analysis lab meets for three hours twice per week with eight students per section taught in the spring term and is preceded by instrumental analysis lecture in the fall. In the laboratory, greater familiarity with a few instruments is acquired rather than experience with numerous instruments. The tradeoff is that the students are more likely to achieve an enhanced understanding of the instruments they do use - learning more with less.

The course is divided into two sequential parts: development and evaluation of instrument standard operating procedures (SOPs) and a real-world research project. Nine weeks are used for SOP developments and evaluations followed by seven weeks for the research project. Students are divided into teams of two or three students. During the SOP phase, each team develops a protocol and determines specifications for an instrument. Subsequently, two other teams evaluate and edit each SOP. Towards the end of the SOP phase, teams write a project research proposal for approval by the instructor. Because SOPs have been written for most instruments, the focus of the student work is on the research problem, not learning a new instrument. The course concludes with written and oral project reports. The primary role of the instructors again is that of a facilitator. Because this is a junior level course, the laboratory assignments are not in a guided inquiry format, but instead, an open ended format is used.

Standard Operating Procedures

Using instrument manuals, students find the information to needed write an SOP for that instrument. Because instrument manuals are the only information provided, most students are at first intimidated with the idea of working on an instrument that they have only seen illustrations of and learned about in lecture. It is not uncommon in commercial and research labs for an employee to be assigned an SOP development for a new instrument with which the employee has never worked.

The SOP must provide enough fundamental details in order for a new user to operate the instrument, run a sample, and obtain a spectrum or chromatogram, as the case may be. Students decide on a simple analyte and use least squares to build a calibration model using a selected wavelength, mass to charge ratio, retention time, etc., depending on the instrument. The figures of merit: precision, accuracy, sensitivity, detection limit, linear dynamic range, and selectivity are determined for the calibration and represent the instrument's specifications. An instrument block diagram is required as well as an analyte spectrum or chromatogram. If it is instrumentally possible, a procedure is included for exporting spectra or chromatograms into Excel.

Students are allowed five weeks to develop their first SOP. This provides ample time to read through manuals and develop a protocol for running the instrument. While no real-world samples are involved, the students do gain the advantages of PBL and CL. Team members work together to agree on what to include and not include in the SOP. Because the team writing the initial SOP has gone through the manuals, these students are considered the resident experts for that instrument. All future questions concerned with the instrument are directed towards them. Student names, phone numbers, and e-mails are listed on the SOP. This team mimics technical service for the instrument manufacturer.

After the initial SOP is written, the SOPs are distributed to other teams for evaluation and editing. In the work place, employees are often required to critique and edit a colleague's report. Two weeks are allowed for teams to review the SOP by making solutions and measuring samples to obtain the calibration curve in the SOP, comparing figures of merit, and assessing the clarity of the SOP. While a reviewing team can go to the manual if they are unsure of a step described in the SOP, this is discouraged. Instead, reviewing teams are expected to contact the authors of the original SOP. At the end of the two week period, teams turn in a rewritten SOP to the instructor, an itemized list of changes, and their evaluation. Copies of the itemized list of changes and evaluation are provided to the original authors of the SOP.

The revised SOPs are distributed to new teams and the students are allowed another two weeks to work through the SOP and revise it. The same tasks described for the first SOP rotation are required for the second. The two SOP rotations are designed to provide students with experience on molecular and atomic spectrometers and a chromatographic system.

After completing the SOPs, students are no longer apprehensive about using a new instrument. Confidence increases and they have developed a sense of independence in the laboratory, which is useful for the project portion of the course. This new found confidence stems from the fact that the students had to figure out how to run an instrument that, for the most part, is alien to them. A fear students overcome is that of breaking the instrument. Because they have ample time to learn the instrument for their first SOP, they soon adapt to trying out buttons and switches with which they initially were not confident from reading the manual. They realize that the instrument is not always going to break when a new feature is explored. Additionally, instruments commonly need repair during the semester. Having students troubleshoot malfunctions provides invaluable experience to boost confidence. In traditional laboratory experiments, students learn that mistakes are usually costly, both in time and with their grade. In the new course, students realize that mistakes are really learning experiences, not something to fear.

Research Project

In the seven-week research project, students solve a problem instead of executing an analysis. The students select a sample type, analyte, and design the analysis using at least one instrument. If pertinent, students can consider during the project why the analysis did not work, e.g., what are the possible sources of error and what can be done to test for the presence of the errors. The project portion of the course concludes with written and oral reports.

Regardless of whether students pursue industrial or academic paths, chances are they will have to prepare a proposal with a budget. To complement the experience in quantitative analysis, students write a proposal for their project that is due towards the end of the SOP portion. It must contain a title, abstract, project narrative with objectives, experimental plan, equipment needed with costs (obtained from the Fisher Scientific catalog for chemicals, etc. and from an instrument manufacture for the chosen instrument), and references. If possible, the official reference method of analysis for the analyte must be included, e.g., American Society for Testing and Materials (ASTM), Environmental Protection Agency (EPA), or the Standard Methods for Examination of Water and Wastewater (*102*). Not all projects end up with a complete analysis and students quickly learn that an analysis of a real-world sample is not straightforward.

Because most students have not had experience writing a formal lab report, a published paper for class discussion is provided during the last week of the SOP phase. The first lab period of the project portion of the course is devoted to understanding the paper and its structure. Students turn in a written overview of the paper and each team is assigned a different section of the paper to orally present. After oral presentations, the class engages in a discussion of the paper.

Topics addressed include conciseness and clarity of the introduction section to introduce subsequent text, completeness of the experimental section, data analysis thoroughness, and if conclusions are warranted. Student oral presentations and discussion of section structure for respective content teaches students the proper way to write the project report. The process also provides practice orally conveying scientific thought.

Oral project reports are made in the last week of school with written reports due at the end of finals week thereby allowing time for teams to edit their reports after instructor and class discussions of their oral reports that follow respective presentations.

Example project titles are: ICP Analysis of Lead in Paint Chips; Analysis of Aspartame, Benzoic Acid, Caffeine, and Saccharine in the Classification of Diet Sodas; Qualitative Analysis of Apple Cultivars with Near-Infrared Spectroscopy; Analysis of Citrus Fruit pH using Diffuse Reflectance Infrared Spectroscopy; Multivariate Analysis of Milk: Use of PCR to Predict Fat Content; Principal Component Analysis of Water Samples; Determination of Fluoride in Toothpaste using Ion Chromatography; Concentration of Contaminants in Soil as a Function of Distance from the Contaminating Source; Determining Iron Content in Foods by UV/Vis Spectrophotometry; and Near IR Classification of Pizza using Principal Component Analysis and Regression.

Chemometrics

With the new instrumental analysis laboratory structure, ample time is available for students to pursue basic and advance chemometric projects. From the quantitative analysis laboratory course, students are familiar with the basics of multivariate calibration by PCR. Some instructor time may be required to explain classification methods for cluster analysis, such as KNN, optimization algorithms, such as genetic algorithms, simplex, or simulated annealing, and experimental design. In one semester, the research project required some use of chemometrics. The introduction section highlighted various chemometric calibration, classification, optimization, and experimental design applications in instrumental analysis have been described. Books and software are available that explain basic material for students to get started with (*103-107*).

Assessment

The described approach has not been formally assessed. However, the teaching methodologies implemented are well documented (*87-90*).

Additionally, this teaching format matches the goals set out in various reports on curriculum issues *(1,2,93,94)*.

Conclusions

A common question often asked by students is, "How is what I am learning now ever going to benefit me in the future?" Students can be heard to say, "I don't need to learn this because I will never need this information in the future." Essentially, undergraduates frequently do not see the relevancy of their course work. This is often the case with traditional general chemistry and quantitative and instrumental analysis laboratories. The basis for the above questions is eliminated when laboratory courses are designed where students realize that the processing skills they learn are actually life-long employment skills. Two primary approaches to incorporating the science process into laboratory courses exist: use an inquiry based method where 1) chemical principles are rediscovered or 2) integrate a real-world application to learn the importance of chemistry. The approaches can be guided or open ended.

With the distinctive approaches described in this chapter, students should be better prepared for the real-world. Students actively participate in the scientific process creating their own knowledge through a discovery based pedagogy and using real-world applications. From this, students should be able to think scientifically with improved reasoning ability to make their own decisions, confidence increases, gain experience in written and oral communication of scientific concepts, and budgetary matters are exemplified. In the quantitative and instrumental analysis courses, students gain experience in the complete process of performing an analysis starting with identifying the problem, collecting samples, sample workup and pretreatment, making measurements, data analysis, and validation of the results.

References

1. Kuwana, T. Curricular Developments in the Analytical Sciences: A Report from NSF Workshops (October 28-30, 1996 and March 13-15, 1997). http://www.chem.ku.edu/TKuwana/CurricularDevelopment/PDF_report.pdf (accessed Sept. 2006).
2. Mabrouk, P.A. *J. Chem. Educ.* **1998**, *75*, 527-528.
3. DePalma, R.A.; Ullman, A.H. *J. Chem. Educ.* **1991**, *68*, 383-384.
4. Thorpe, T.M. *J. Chem. Educ.* **1986**, *63*, 237.
5. Thorpe, T.M.; Ullman, A.H. *Anal. Chem.* **1996**, *68*, 477A-480A.

6. Kratochvil, B.J. *J. Chem. Educ.* **1991**, *68*, 838-839.
7. McKinnon, I.R.; Nunn, E.K.; Patti, A.F. What laboratory-related skills do employers want from chemistry graduates? *Proceedings of the Royal Australian Chemical Institute Chemical Education Division Conference*, University of Central Queensland: Rockhampton, Australia, July 1998; pp 155-156.
8. Beck II, C.M. *Anal. Chem.* **1991**, *63*, 993A-1003A.
9. Spencer, J.N. *J. Chem. Educ.*, **1999**, *76*, 566-569.
10. Hanson, D.; Wolfskill, T. *J. Chem. Educ.*, **2000**, *77*, 120-130.
11. The POGIL Project. Process Oriented Guided Inquiry Learning. www.pogil.org (accessed Sept. 2006).
12. Hanson, D. *Instructor's Guide to Process-Oriented Guided-Inquiry Learning*; Pacific Crest, Lisle, IL; 2006.
13. Kovacs, J. *J. Chem. Educ.*, **1999**, *76*, 120-124.
14. Ross, M.R.; Fulton, R.B. *J. Chem. Educ.*, **1994**, *71*, 141-143.
15. Karukstis, K. *J. Chem. Educ.*, **2005**, *82*, 1440-1441.
16. Karukstis, K.K.; Wenzel, T.J. *J. Chem. Educ.*, **2004**, *81*, 468-469.
17. Hanks, T.W.; Wright, L.L. *J. Chem. Educ.*, **2002**, *79*, 1127-1130.
18. Wenzel, T.J. *Anal. Chem.* **1995**, *67*, 470A-475A.
19. Wright, J.C. *J. Chem. Educ.* **1996**, *73*, 827-832.
20. Walters, J.P. *Anal. Chem.* **1991**, *63*, 1179A-1191A.
21. Wilson, G.S.; Anderson, M.R.; Lunte, C.E. *Anal. Chem.* **1999**, *71*, 677A-681A.
22. Hope, W.W.; Johnson, L.P. *Anal. Chem.* **2000**, *72*, 460A-467A.
23. Wener, T.C.; Tobiessen, P.; Lou, K. *Anal. Chem.* **2001**, *73*, 84A-87A.
24. Phlips, D.N. *Anal. Chem.* **2002**, *72*, 427A-430A.
25. Kalivas, J.H. *J. Chem. Educ.* **2005**, *82*, 895-897.
26. Hughes, K.D. *Anal. Chem.* **1993**, *65*, 863A-889A.
27. Houghton, T.P.; Kalivas, J.H. *J. Chem. Educ.* **2000**, *77*, 1314-1318.
28. Wenzel, T.J. *Anal. Chem.* **1999**, *71*, 693A-695A.
29. Wenzel, T.J. *Anal. Chem.* **1998**, *70*, 790A-795A.
30. Wenzel, T.J. *Anal. Chem.* **2000**, *72*, 293A-296A.
31. Wenzel, T.J. *Anal. Chem.* **2000**, *72*, 359A-361A.
32. Towns, M.H. *J. Chem. Educ.* **1998**, *75*, 67-69.
33. Öberg, T. *J. Chem. Educ.* **2006**, *83*, 1178-1181.
34. Delaney, M.F.; Warren, jr, F.V. *J. Chem. Educ.* **1981**, *58*, 646-651.
35. Howery, D.G.; Hirsch, R.F. *J. Chem. Educ.* **1983**, *60*, 656-659.
36. auf der Hyde, T.P.E. *J. Chem. Educ.* **1990**, *67*, 461-469.
37. Chau, F.T.; Chung, W.H. *J. Chem. Educ.* **1995**, *72*, A84-A85.
38. Msimanga, H.Z.; Elkins, P.; Tata, S.K.; Smith, D.R. *J. Chem. Educ.* **2005**, *82*, 415-424.

39. Raymond, M.; Jochum, C.; Kowalski, B.R.; *J. Chem. Educ.* **1983**, *60*, 1072.
40. Blanco, M.; Iturriaga, H.; Maspoch, S.; Tarín, P. *J. Chem. Educ.* **1989**, *66*, 178-180.
41. Pandey, S.; McHale, M.E.R.; Coym, K.S.; Acree, Jr., W.E. *J. Chem. Educ.* **1998**, *75*, 878-880.
42. Ribone, M.E.; Pagani, A.P.; Olivieri, A.C.; Goicoechea, H.C. *J. Chem. Educ.* **2000**, *77*, 1330-1333.
43. Cartwright, H. *J. Chem. Educ.* **1986**, *63*, 984.
44. Harvey, D.T.; Bowman, A. *J. Chem. Educ.* **1990**, *67*, 470-472.
45. Charles, M.J.; Martin, N.W., Msimanga, H.Z. *J. Chem. Educ.* **1997**, *74*, 1114-1117.
46. Rusak, D.A.; Brown, L.M.; Martin, S.D. *J. Chem. Educ.* **2003**, *80*, 541-543.
47. Cazar, R.A. *J. Chem. Educ.* **2003**, *80*, 1026-1029.
48. Horovitz, O.; Sârbu, C. *J. Chem. Educ.* **2005**, *82*, 473-483.
49. Curtis, M.A. *J. Chem. Educ.* **1994**, *71*, 775-778.
50. Sangsila, S. Labinaz, G.; Poland, J.S.; van Loon, G.W. *J. Chem. Educ.* **1989**, *66*, 351-353.
51. Stolzberg, R.J. *J. Chem. Educ.* **1999**, *76*, 834-838.
52. Amenta, D.S.; Lamb, C.E.; Leary, J.J. *J. Chem. Educ.* **1979**, *56*, 557-558.
53. Leggett, D.J. *J. Chem. Educ.* **1983**, *60*, 707-710.
54. Shavers, C.L.; Parsons, M.L.; Deming, S.N. *J. Chem. Educ.* **1979**, *56*, 307-309.
55. Steig, S. *J. Chem. Educ.* **1986**, *63*, 547-548.
56. Meftah, M.E.; Bigan, M.; Blondeau, D. *Chem. Educ.* **2003**, *8*, 318-326.
57. Palasota, J.A.; Deming, S.N. *J. Chem. Educ.* **1992**, 69, 560-563.
58. Capello, G.; Goicoechea, H.C.; Miglietta, H.F.; Mantovani, V.E. *Chem. Educ.* **2003**, *8*, 371-374.
59. Gozálvez, J.M.; García-Díaz, J.C. *J. Chem. Educ.* **2006**, *83*, 647.
60. Oles, P.J. *J. Chem. Educ.* **1998**, *75*, 357-359.
61. van Ryswyk, H.; van Hecke, G.R. *J. Chem. Educ.* **1991**, *68*, 878-882.
62. Harrington, P.B.; Kolbrich, E.; Cline, J. *J. Chem. Educ.* **2002**, *79*, 863-864.
63. Lang, P.L.; Miller, B.I.; Nowak, A.T. *J. Chem. Educ.* **2006**, *83*, 280-282.
64. Jeannot, M.A. *J. Chem. Educ.* **2006**, *83*, 243-244.
65. Bell, S.C.; Moore, J. *J. Chem. Educ.* **1998**, *75*, 874-877.
66. Marcos, Ríos, A.; Valcárcel, M. *J. Chem. Educ.* **1995**, *72*, 947-949.
67. Laquer, F.C. *J. Chem. Educ.* **1990**, *67*, 900-902.
68. Kalivas, J.H.; Blount, C.W. *J. Chem. Educ.* **1988**, *65*, 794-795.
69. Ditzler, M.A.; Ricci, R.W. *J. Chem. Educ.*, **1994**, *71*, 685-688.
70. Lamba, R.; Sharma, S.; Loyd, B.W. *J. Chem. Educ.*, **1997**, *74*, 1095-1099.
71. Farrell, J.J.; Moog, R.S.; Spencer, J.N. *J. Chem. Educ.*, **1999**, *76*, 570-574.

72. Lewis, S.E.; Lewis, J.E. *J. Chem. Educ.*, **2005**, *82*, 135-139.
73. Lechtanski, V.L. *Inquiry-Based Experiments in Chemistry*; American Chemical Society, Oxford University Press; 2000.
74. Walczak, M.M.; Lantz, J.M. *J. Coll. Sci. Teaching* **1998**, *28*, 152-157.
75. Lantz, J.M.; Feindt, J.E.; Lewellyn, E.P.B.; Walczak, M.M. *J. Chem. Educ.* **1999**, *76*, 1671-1672.
76. Walczak, M.M.; Lantz, J.M. *J. Chem. Educ.* **2004**, *81*, 218-220.
77. Peaslee, G.; Lantz, J.M.; Walczak, M.M. *J. Coll. Sci. Teaching* **1998**, *28*, 21-24.
78. Lantz, J.M.; Walczak, M.M. *Chem. Educator* **1997**, *1*, S 1430-1471.
79. O'Hara, P.B.; Sanborn, J.A., Howard, M. *J. Chem. Educ.*, **1999**, *76*, 1673-1677.
80. Kesner, L.; Ewing, E.M. *J. Chem. Educ.* **1999**, *76*, 920-923.
81. Wiegand, D.; Strait, M. *J. Chem. Educ.* **2000**, *77*, 1538-1539.
82. Hatch-Seekers, M.; Aragon, E. *J. Chem. Educ.* **2002**, *79*, 462-464.
83. Draper, A. *J. Chem. Educ.* **2004**, *81*, 221-224.
84. Esson, J.M.; Stevens-Truss, R.; Thomas, A. *J. Chem. Educ.* **2005**, *82*, 1168-1173.
85. Kalivas, J.H., Idaho State University, *unpublished.*
86. http://www.wcer.wisc.edu/salgains/instructor/ (accessed Sept. 2006).
87. Johnson, D.W.; Johnson, R.T; Smith, K.A. *Active Learning: Cooperation in the College Classroom*; Interaction Book Company: Edina, MN, 1991.
88. *How People Learn: Brain, Mind, Experience, and School*; Bransford, J.D.; Brown, A.L.; Cocking, R.R., Eds.; National Academy Press: Wahsington D.C., 1999.
89. *Chemist's Guide to Effective Teaching*; Pienta, N.J.; Cooper, M.M.; Greenbowe, T.J., Eds.; Pearson Prentice Hall: Upper Saddle River, NJ, 2005.
90. Heuvelen, A.V. *Amer. J. Physics* **1991**, *59*, 891-897.
91. Johnston, A.H. *J. Chem. Educ.* **1997**, *74*, 262-268.
92. Spence, J.N. *J. Chem. Educ.* **2006**, *83*, 528-533.
93. Wenzel, T.J. Enhancing Research in the Chemical Sciences at Predominately Undergraduate Institutations: A Report from NSF Workshops (August 2-4, 2003). http://abacus.bates.edu/acad/depts/chemistry/twenzel/summit.html (accessed Sept. 2006).
94. *Reinvigorating the Undergradaute Experience: Successful Models Supported by NSF's AIRE/RAIRE Program*; Kauffman, L.; Stocks, J., Eds.; CUR: Washington D.C., 2004. http://www.cur.org/publications/aire_raire/toc.asp (accessed Sept. 2006).

95. *Official Methods of Analysis of AOAC International*; Cunniff, P., Ed.; AOAC International: Gaithersburg, MD, 1997; Vols I and II.
96. Harris, T.M. *J. Chem. Educ.* **1993**, *70*, 340-341.
97. Sherren, A.T. *J. Chem. Educ.* **1991**, *68*, 598-599.
98. Gilles de Pelichy, L.D.; Adams, C.; Smith, T. *J. Chem. Educ.* **2005**, *82*, 1192-1194.
99. Calascibetta, F.; Campanella, L.; Favero, G. *J. Chem. Educ.* **2000**, *77*, 1311-1313.
100. Mathias, J.A.; Williams, P.C.; Sobering, D.C. *Aquaculture* **1987**, *61*, 303-311.
101. Harris, D.C. *Quantitative Chemical Analysis*; W.H. Freeman and Company: New York, NY, 2007; pp 39-95.
102. *Standard Methods for Examination of Water and Wastewater*, 19th Edition; American Public Health Association: Washington DC, 1975.
103. Beebe, K.R.; Pell, R.J.; Seaholtz, M.B. *Chemometrics: A Practical Guide*; Wiley: New York, NY, 1998.
104. Næs, T.; Isaksson, T.; Fearn, T.; Davies, T. *A User-Friendly Guide to Multivariate Calibration and Classification*; NIR Publication: Chichester, UK, 2002.
105. Wise, B.M.; Gallagher, N.B.; Bro, R.; Shaver, J.M. *PLS_Toolbox 3.0 for use with MATLAB*, Eigenvector Research: Manson, WA, 2003.
106. Olivieri, A.C.; Goicoechea, C.; Iñón, F.A. *Chemom. Intell. Lab. Syst.* **2004**, *73*, 189-197.
107. *Practical Guide to Chemometrics*; Gemperline, P., Ed.; CRC Press Taylor & Francis: Boca Raton, FL, 2006.

Chapter 12

Web-Based Applications

Effective Application of Spectroscopy and Chemistry Software Tools within the Analytical Chemistry Curriculum to Facilitate Learning and Labwork

Marie Scandone, Deborah Kernan, and Gregory M. Banik

Informatics Division, Bio-Rad Laboratories, Inc., 3316 Spring Garden Street, Philadelphia, PA 19104

Increasingly, teachers are incorporating the use of chemistry software programs within their departments and curriculum. The benefits of doing so are enormous and can accelerate the learning process. Chemistry software can allow students to understand the concepts within the curriculum by visualization as well as by theory. These tools also allow students to present the results of their research in a clear and concise manner. Finally, real experience with informatics tools can aid in academic research as well as establish a pathway to industrial applications. The use of chemistry software in the classroom can specifically help teachers in the instruction of fundamentals of structure, bonding and spectroscopy, and can facilitate the development of skills from basic analytical skills to high-order critical thinking. It can also help teachers to motivate their students and allows them to work at their own pace, not unlike e-learning. Overall, the incorporation of such tools into the curriculum provides more possibilities and flexibility in both the teaching and learning process. There are a number of software programs available for academia and this chapter will highlight the most useful for chemistry teachers in the area of structure drawing, visualization, general chemistry, and spectroscopy. It will also give a classroom example of how the software can be incorporated into the curriculum.

Structure Drawing and Visualization Software

Presently, there are a number of software programs that are considered all-purpose chemical drawing and graphics software. Some are freeware for academic use. They have been widely reviewed and tested. They include ChemSketch, ISIS\Draw, ChemDraw, Chemistry 4-D Draw and ChemWindow® (KnowItAll® Academic Edition). These are the major software packages and there are comparisons available in print and on the web that highlight their features.

ACD - ChemSketch

Advanced Chemistry Development, Inc. (ACD/Labs) offers one such product to the chemical community. They offer a version of this chemical drawing package on-line (ACD/ChemSketch). The software provides a drawing interface with a portal to analytical tools and allows the transformation of structural or analytical data into reports or presentations.

It also includes tautomer recognition, 2D structure cleaning, 3D optimization and viewing, and drawing of polymers, organometallics, Markush structures (generic view) and special Markush structures with added or removed mass or fragments to describe metabolic and mass-spectral transformations, and structures with delocalization. The Chemistry features allow the display of chemical formulas, molecular weights, percentage composition, and estimated macroscopic properties: molar refractivity, refractive index, molar volume, density, parachor, and others.

Finally, this package permits the user to create chemistry-related reports and presentations. It offers academic institutions one of many software options for interactive learning. The software is available at http://www.acdlabs.com/download/chemsk.html as well as information on its use.

CambridgeSoft - ChemDraw

CambridgeSoft's ChemDraw is a structure drawing and molecular modeling software. It facilitates the creation of structures from chemical names and produces IUPAC names for structures. It permits the estimation of an NMR spectrum from a *ChemDraw* structure with direct atom to spectral correlation. With this software, database queries can be created by specifying atom and bond properties, and it includes stereochemistry. ChemDraw enables the display of spectra, structures, and annotations and can be used for many types of compounds (including charged compounds and salts, highly symmetric structures, and many other types of inorganic and organometallics).

Access to a demo version of this software as well as other modules is available at http://www.cambridgesoft.com/software/ChemDraw/.

ChemInnovation - Chemistry 4-D Draw

ChemInnovation's Chemistry 4-D Draw is a chemistry program for structure drawing and more. With this software, the user can create high-quality structures by entering molecular names or do the reverse and assign systematic names to structures. It includes a full set of tools for drawing, text, and structure editing, and labeling. Other features include interactive 3-D rotation, syntax checking, hot-key labeling, multi-step undo, and the ability to create structure templates with user-defined trivial names. The software enables the user to assign chirality and to perform object transformations such as rotation, movement, scaling, flipping, and aligning. Fonts and colors can be customized. Files generated can be exported as MOL files and graphic files.

Access to this software as well as other modules is currently available as a trial download at http://www.cheminnovation.com/products/chem4d.asp.

MDL® - ISIS/Draw

Elsevier's MDL® ISIS/Draw is a drawing package that enables the drawing of chemical structures using the signs and symbols used in paper sketches. It allows the insertion of ISIS/Draw sketches into documents, web pages, spreadsheets, and presentations. It can be used to create structures to register into 2D and 3D molecule, polymer, and reaction databases, and to create queries for searching these databases.

The main features of this software are the ability to create chemical structures, including biomolecules and polymers, the ability to control the look of the structures with scaling of atoms and bonds, the ability to customize fonts and other attributes.

This software can be used to post structures on the web, as well as integrate with other software. Users can cut and paste ISIS/Draw sketches into Microsoft Word, Excel, PowerPoint, and other software, or, take advantage of OLE on Windows.

Access is available through the MDLI website at http://www.mdli.com/. [UK mirror http://www.mdli.co.uk.]

Bio-Rad - KnowItAll® Academic Edition

Bio-Rad Laboratories, Inc., Informatics Division, currently provides a completely free and fully functional software package to promote learning and research in the academic community. All students, faculty, and staff of any degree-granting academic institution worldwide are eligible to use this software at no charge to the student or institution.

This software combines structure drawing software tools with spectroscopy software tools all with one interface. With the KnowItAll® Informatics System, Academic Edition, students can draw structures; perform IR and Raman functional group analysis; import, process, and interpret spectra; calculate molecular weight, point groups, bond lengths, and angles; convert mole to mass; perform MS fragmentation; digitally access Sadtler's spectral handbooks and generate reports. Since all of these tools reside in one common interface, the KnowItAll applications are easy to learn and hence quite easy to incorporate and introduce into the classroom. These tools can be effectively incorporated into any chemistry curriculum, including general, organic and analytical chemistry classes. Teachers involved in spectroscopy, the interpretation of spectral data, and analytical coursework will find this software valuable. In this chapter, we will give a brief overview of each application's functionality, demonstrate how these tools can be utilized, and provide specific examples that can be used in the classroom for each tool.

Other tools available for learning are the tutorials or training movies which can instruct teachers and students to use the software and get the most out of the applications. These training resources can save valuable class time by allowing the students to learn at their own pace and become proficient in the use of these tools.

Bio-Rad offers single-user copies to students and teachers or site licenses for entire departments or schools if they meet the eligibility requirements outlined in the Terms and Conditions of Use Agreement. To download the software, go to www.knowitall.com/academic.

The following applications and features are included in the KnowItAll Academic Edition:

DrawIt™ for Chemical Structure Drawing

An essential component of any chemistry software is a comprehensive set of chemistry drawing software tools designed for the chemist who needs to produce chemical structures. Communication of all results is essential and chemical structures can be used to identify results. The user should have the ability to draw any chemical structure with little difficulty. KnowItAll's DrawIt software application permits the user to store fully integrated chemical structure fields and use pre-designed tables and forms to enter chemical data. Users can draw any chemical structure with just a few clicks and drags. It includes all the tools needed to draw rings, bonds, atoms, chains, arrows, and chemical symbols. With programmable hot keys, teachers and students can quickly label common groups by typing a single key or bringing in a partial or common structure they use often with a single click.

Students can use this tool to present clear representations of their work and learn chemical bonding theory at the same time. The software's Chemistry Checker, which works like a spell checker in a word processing program, but for chemical structures, is useful to those learning how to draw chemical structures. The checker recognizes many types of problems including incorrect valences in connected bonds, incorrect valences in atoms labels and characters in a label which no chemical meaning. When a problem is found, the location of the suspected error is circled and a brief description of the problem appears for the user. This permits the user to then redraw the structure in the correct manner. As such, it is an excellent tool that shows the students how to understand the chemistry behind the structure.

ReportIt™ to Generate Lab Reports

Another essential component of any chemistry software is publishing software designed for the chemist who needs to produce professional reports, chemical structures, chemical reactions, lab experiment setups, chemical engineering diagrams, data tables, and other information. The software also permits the import of IR, NMR, MS and Raman spectra and chromatograms in common native formats; it is not instrument- or vendor-dependent and can be used by all spectroscopy laboratories that produce data, no matter what technique is used.

This software is also useful for those who need to illustrate experiments and engineering processes. Teachers and students can use this application to construct high-quality lab reports and ensure that their work is clearly and precisely presented. The software also includes a Laboratory Glassware Collection contains more than 130 illustrations that are critical to the description of a laboratory experiment. All pieces are drawn to scale and snap together at joints for easy construction and quick reporting. The second collection included – the Chemical Engineering Collection helps students and teachers to draw realistic process flow diagrams; it offers more than 250 Process Flow Symbols including furnaces, filters, compressors, coolers, exchangers, etc.

Reports are easy to lay out by using one of many predefined templates or the student/teacher may opt to create their own. The reports can then be transferred to PowerPoint slides to allow the sharing of these results.

AnalyzeIt™ IR & Raman to Interpret Spectra

Analysis is often done on materials that are not readily available in spectral databases or cannot be interpreted by the user. Once these substances are synthesized, and to advance the development process, identification, or at the very least, classification is essential. Therefore, this interactive software assists in the interpretation of infrared and Raman spectra. The application contains a database of characteristic group frequency spectral ranges and intensities for each technique. The results will display all of the functional groups in the database that contain a peak which falls within the tolerance set for the peak selected. The software will also list the type of chemical bond that causes the absorption, any other characteristic absorption bands of that functional group, normal relative intensity for each band of the specified functional group and the substructure for that functional group. It can provide clear and rapid verification and identification by providing information on functional groups. The knowledge base not only aids experienced analysts in interpreting complex spectra, but it can also be used to instruct novice IR spectroscopy users in the basics of spectral interpretation. This makes AnalyzeIt useful for teaching the fundamentals of functional group analysis without the need to consult the numerous reference sources on the subject. Interpretation of Infrared and Raman spectra can directly be accomplished with the software and students can learn the principles of symmetry and bonding.

To interpret the bands in an Infrared or Raman spectrum, the student first opens an experimental spectrum and then clicks on a peak of interest to generate a list of all functional groups possible at that frequency. AnalyzeIt's knowledgebase includes over 200 functional groups and hundreds of interpretation frequencies. It also permits the correlation from a structure and helps to determine whether a structure matches an infrared or Raman spectrum. Simply draw a structure and AnalyzeIt will break the structure into its component functional groups to overlay with the spectrum.

RefineIt™ IR and RefineIt™ Raman

The RefineIt IR and Raman applications provide a variety of tools useful in processing spectral data such as Baseline Correction, Flatline, Smoothing, Truncation and Padding, Normalization, ATR Correction, Kubelka-Monk calculations, Spectral Addition/ Subtraction and Peak Picking. RefineIt can be used to improve the accuracy of searches as well as the overall appearance of reports.

ProcessIt™ NMR

NMR spectrometers record data in the free-induction decay (FID) or time-domain form. Fourier transform (FT) of the FID converts it to the familiar NMR spectrum (frequency domain data) with frequency in Hz or ppm displayed on the x-axis and intensity on the y-axis. The conversion from FID as well as operations used to improve appearance are referred to as spectrum processing.

With ProcessIt NMR, users can seamlessly import and process NMR spectra from various sources. This software offers a comprehensive set of processing features to correct experimental artifacts and improve the appearance of spectra. It also saves valuable processor time at the instrument, thereby improving sample throughput. The tools include Baseline Correction, Phasing, Peak Picking, Set Reference, Integration, Addition\Subtraction, and Processing FID to FT or FT to FID.

SymApps™ for Molecular Symmetry Analysis

This application is a professional symmetry analysis and 3-D molecular rendering program, designed for desktop visualization and publishing. A modified MM2 force field minimization module converts 2-D structure drawings to 3-D in just seconds. It calculates, displays, and animates the symmetry for a molecule including rotation axis, mirror planes, and inversion centers. Students can also use this application to create movies for three basic rotations and export them as .avi files that will run in any application that supports this format. The software also allows student to calculate and understand crucial information and concepts such as point groups, bond lengths, angles, and dihedral angles for all atoms in a structure.

General Laboratory Calculations

The software calculates mass from a chemical structure and has easy mole-to-mass conversion included. Another feature is the MS Fragmentation tool which offers an efficient means to determine whether a proposed structure matches mass spectral data. This tool draws a movable fragmentation line through the proposed structure, displays the formula and mass for the fragments on either side of the line, and displays the fragment of the structure. Another MS tool has been designed specifically for documentation. This tool creates a fragmentation line and displays the masses of the two fragments. Students can use these calculations to check their data and confirm the MS fragmentation patterns in their structures.

IQ Academic Spectral Database

Another tool that can be acquired from Bio-Rad as a supplement to the tools in the KnowItAll Academic Edition is the IQ Academic Spectral Database specifically designed for teaching. This database offers teachers and students a convenient collection of Infrared and NMR spectra of organic compounds relevant to college introductory courses on organic chemistry and the supplementary laboratory courses on experimental organic chemistry and qualitative organic analysis. It can be used effectively by teachers to prepare examinations and test the students on the fundamentals using spectral data. In addition to serving as a teaching aid, this database, which contains 1,279 compounds with over 3,800 spectra, can used by students to establish the identity of a compound in the laboratory through empirical comparison via functional group analysis. This archive of multi-technique information can be used as a collection of spectral data, a complementary database to any spectroscopy course, and an archive of multi-technique information.

This teaching tool offers easy access to high-quality spectral data and presents multi-technique spectral data (IR, HNMR and CNMR) for each compound in the database. Compounds are identified by name, CAS Registry Number, and structure. Properties such as molecular formula, molecular weight, melting point, and or boiling point are listed when available, as well as the origin and parameters for the NMR spectral data. Each record also contains a web link to an entry in the online version of the well known IR and NMR Sadtler Handbooks for that technique.

The database includes specially selected infrared spectra that cover a broad range of chemical classes useful within the chemistry curriculum. A number of these spectra are also cross-referenced to CNMR and HNMR spectra and were provided by the National Institute of Advanced Industrial Science and Technology, and NMRDBTech, Inc.

The IQ Database can be used by teachers to prepare examinations and test the students on the fundamentals using spectral data. It can be used by teachers to create problems and assignments that reinforce the fundamentals of science. Spectroscopy problems can be broken down into two simple categories: Compound Verification and Unknown Identification and two questions asked are "Do I have what I think I have" or "What do I have"? From IR and Raman we learn which functional groups are present and from NMR, we learn how carbon and hydrogen atoms are bonded.

Use in the Classroom

With this software program, teachers can use the spectral database as well as all the applications in the package to provide their students with experience of working in a spectral laboratory. In the following example, the students will be asked to use spectral data to identify "unknown" chemical compounds. The student is asked to make an identification of a saturated, aliphatic hydrocarbon. It is a colorless liquid with a gasoline-like odor. In the infrared, all straight-chain hydrocarbons exhibit the same vibrations. There are five major areas of similarity.

1. C-H stretching vibration:
   - CH_3 asymmetric stretching, 2972-2952 cm^{-1}
   - CH_3 symmetric stretching, 2882-2862 cm^{-1}
   - CH_2 asymmetric stretching, 2936-2916 cm^{-1}
   - CH_2 symmetric stretching, 2863-2843 cm^{-1}
2. C-H bending vibration:
   - CH_3 asymmetric bending, 1470-1430 cm^{-1}
   - CH_2 asymmetric bending, 1485-1445 cm^{-1} (overlaps band due to CH_3 asymmetric bending)
3. C-H bending vibration:
   - CH_3 symmetric bending, 1380-1365 cm^{-1} (when CH_3 is attached to a C atom)
4. C-H wagging vibration:
   - CH_2 out-of-plane deformations wagging, 1307-1303 cm^{-1} (weak)

5. CH_2 rocking vibration:

   $(CH_2)2$ in-plane deformations rocking, 750-740 cm^{-1}
   $(CH_2)3$ in-plane deformations rocking, 740-730 cm^{-1}
   $(CH_2)4$ in-plane deformations rocking, 730-725 cm^{-1}
   $(CH_2) \geq 6$ in-plane deformations rocking, 722 cm^{-1}

Splitting of the absorption band occurs in most cases (730 and 720 cm^{-1}) when the long carbon-chain alkane is in the crystalline state (orthorombic or monoclinic form).

Therefore, it would be difficult to make a precise identification from the infrared spectrum alone. However, if the proton NMR and the Carbon-13 NMR were available, this additional information would make the identification easier. With the peak information provided, the student can ascertain that the unknown was symmetrical and with the integration of the proton NMR spectrum; they can also identify the end group as a methyl group (CH_3) attached to an ethyl group (CH_2) attached to a CH_2 group. So, the compound is most likely pentane.

Without consulting books or performing experiments, the students are able to understand the principles of spectral interpretation, functional group analysis, molecular structures and formulas as well as IR, HNMR and CNMR spectroscopy.

General Chemistry Software

Besides these software programs, there a number of useful software aids available to teachers and students alike. Below is a sampling of the available tools that teachers can use to facilitate learning and labwork within the analytical chemistry curriculum.

AcidBaseLab is a titration curve calculation program that can be used to simulate real titration experiments. This freeware does more than simply calculating the pH of a strong or a weak acid (or base). It can perform any pH-calculation, independent of the complexity of the composition under consideration. During titration, the color (changes) of the solution can be monitored. Moreover, it enables the user to edit the list of possible acid/base chemicals as well as the pH-indicator list and can match specific needs. It is available at http://chemometrix.ua.ac.be/dl/acidbase/

ArgusLab (molecular modeling, graphics, and drug design program). It is available at http://www.arguslab.com/

Rasmol is available at http://www.umass.edu/microbio/rasmol/index2.html. This software is designed for 3-D viewing protein structures and can be used for small molecules structures.

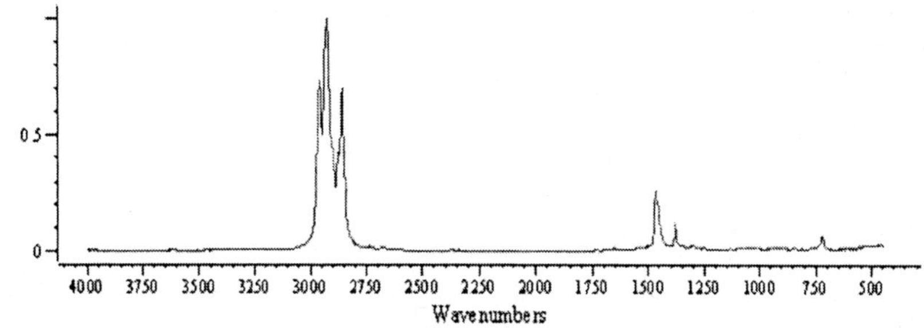

Figure 1. Infrared spectrum (Copyright 2004, Bio-Rad Laboratories, Inc., Informatics Division.)

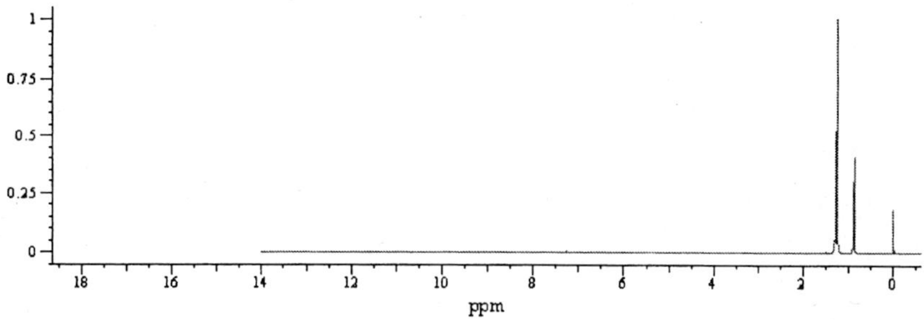

Figure 2. HNMR Spectrum. (Copyright 2004 National Institute of Advanced Industrial Science and Technology and NMRDB Tech. Inc., Japan.)

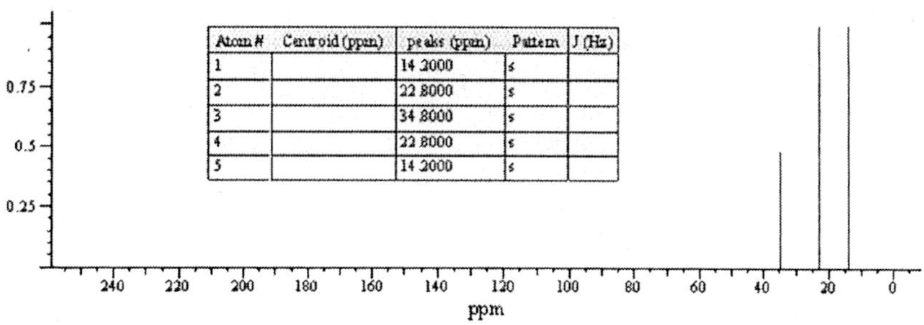

Figure 3. CNMR Spectrum. (Copyright 2004 National Institute of Advanced Industrial Science and Technology and NMRDB Tech. Inc., Japan.)

CHIME (3-D viewer plug-in for web browsers) is available at http://www.mdl.com/products/framework/chime/. CHIME is needed for viewing many chemical and biochemical web sites. It is based on the stand-alone program Rasmol. Chime is a browser plug-in that renders 2D and 3D molecules directly within a Web page. The molecules are "live", meaning they are not just static pictures, but chemical structures that scientists can rotate, reformat, and save in various file formats for use in modeling or database applications.

Maestro (molecular modeling and visualization program) is available from Schrodinger at www.schrodinger.com. It enables the user to build and modify a wide range of chemical structures, view them in a variety of rendering styles, and output high-quality images for use in print and/or presentations. It can be used as an electronic notebook to record important properties associated with each structure.

SweetMollyGrace is a suite of freeware and shareware tools for automating the work of rendering and animating molecules. It imports molecules in PDB, MOL, and XYZ format and generates high quality images in raytracing (Povray and Raster3D). It can import and manage POV (and VRML) files. With this software, the user can create AVI, MPEG, GIF, MOV and FLIC animations and pdb trajectory files, which can be converted to animation. The user can generate and view 3D files in different formats: VRML, DXF, 3DS, OBJ etc., as well as create and view Postscript images. The software is available at http://rodomontano.altervista.org/engSweetMG.php.

The Chemistry Collective is a collection of virtual labs, scenario-based learning activities, and concepts tests which can be incorporated into a variety of teaching approaches as pre-labs, alternatives to textbook homework, and in-class activities for individuals or teams. It is organized for college and high school teachers who are interested in using, assessing, and/or creating engaging online activities for chemistry education. It can be accessed and viewed at http://www.chemcollective.org/.

The Chemical Thesaurus is a subset of the reaction chemistry database, including a database for atoms, isotopes and ions, main group chemistry, organic functional group chemistry, reaction mechanisms, Lewis acids and Lewis bases, redox agents, radicals, VSEPR geometries, and radioactive decay series. The software uses a relational database to store information about any chemical reaction process, including: radiochemistry, phase change, resonance structure interconversion, interchanging conformation, single electron transfer, complexation, substitution-displacement, redox reactions, photochemistry, rearrangements, multi-step mechanisms, synthetic pathways, biosynthetic pathways, and many more. Download it at http://www.chemthes.com/.

Orbital Viewer (draws and calculates atomic orbitals) available at http://www.orbitals.com/orb/ov.htm. It has many features, and comes in both a Windows version and a command-line interface version. It permits the drawing of any atom, and any molecule, and creates animations, cutaways, and shows the locations where the probability goes to zero. It permits the user to light the orbital from any location, cast shadows, and save files in TIFF, PPM, BMP, AVI, and VRML formats.

EniG. Periodic Table of the Elements contains the basic data about the elements in seven languages (English, Croatian, French, German, Italian, Spanish, and Portuguese). The Periodic Table can be incorporated into presentations. The color of background can be adjusted to match the color of the presentation slide. It is available at http://www.ktf-split.hr/~eni/toys/pse-e.html.

Summary

Many of these software tools are excellent – and many of them free of charge additions to any general chemistry, organic chemistry, and analytical chemistry courses in college as well as in high school. They allow students an opportunity to work with informatics software and obtain experience with the tools used everyday in industry. These packages present pertinent teaching aids for experience with real-world problems and tools to solve them.

References

1. Gunda, T.E. Chemical Drawing Programs - The Comparison of ISIS/Draw, ChemDraw, DrawIt (ChemWindow), ACD/ChemSketch and Chemistry 4-D Draw. http://dragon.klte.hu/~gundat/rajzprogramok/dprog.html (accessed Dec 1, 2006), University of Debrecen, POB 36, H-4010 Debrecen, Hungary.
2. Zhenjiang Li, Honggui Wan, Yuhu Shi, and Pingkai Ouyang. *J. Chem. Inf. Comput. Sci.*, **2004**, 44, No. 5 COMPUTER SOFTWARE REVIEWS.
3. Advanced Chemistry Development, Inc. ACD/ChemSketch. http://www.acdlabs.com/ (accessed Dec 1, 2006), Toronto, Canada.
4. MDL Information Systems, Inc. ("Elsevier MDL"). MDL® ISIS/Draw. http://www.mdli.com/ (accessed Dec 1, 2006), San Ramon, CA.
5. CambridgeSoft Corporation. ChemDraw. http://www.camsoft.com/ (accessed Dec 1, 2006), Cambridge, MA,

6. ChemInnovation Software, Inc. Chemistry 4-D Draw. http://www.cheminnovation.com/ (accessed Dec 1, 2006), San Diego, CA.
7. Bio-Rad Laboratories, Inc. KnowItAll® Academic Edition. http://www.knowitall.com/academic/ (accessed Dec 1, 2006), Philadelphia, PA.
8. AcidBaseLab. http://chemometrix.ua.ac.be/dl/acidbase/ (accessed Dec 1, 2006).
9. ArgusLab. http://www.arguslab.com/ (accessed Dec 1, 2006).
10. Rasmol. http://www.umass.edu/microbio/rasmol/index2.html (accessed Dec 1, 2006).
11. MDL Information Systems, Inc. ("Elsevier MDL"). MDL® CHIME http://www.mdl.com/products/framework/chime/ (accessed Dec 1, 2006), San Ramon, CA.
12. Schrodinger, LLC. Maestro. http://www.schrodinger.com (accessed Dec 1, 2006), New York, NY.
13. SweetMollyGrace. http://rodomontano.altervista.org/engSweetMG.php (accessed Dec 1, 2006).
14. The Chemistry Collective. http://www.chemcollective.org/ (accessed Dec 1, 2006).
15. The Chemical Thesaurus. http://www.chemthes.com/ (accessed Dec 1, 2006).
16. Orbital Viewer. http://www.orbitals.com/orb/ov.htm (accessed Dec 1, 2006).
17. EniG. http://www.ktf-split.hr/~eni/toys/pse-e.html (accessed Dec 1, 2006).

Note that this information in this chapter represents current offerings as they exist in the software packages from select vendors at the time the article was written. Offerings may change or may have already changed subsequent to the writing of this article. Please check directly on individual vendors websites for the most up-to-date information and offerings for academia.

The authors make no representation or warranty as to the accuracy or completeness of information presented in this review. Readers are best advised to contact the company(-ies) listed (and others, which may not be listed herein) or visit their websites.

Chapter 13

Active Learning Using the Virtual Mass Spectrometry Laboratory

Mark E. Bier[1] and Joseph J. Grabowski[2]

[1]Department of Chemistry, Center for Molecular Analysis, Carnegie Mellon University, 4400 Fifth Avenue, Pittsburgh, PA 15213
[2]Department of Chemistry, University of Pittsburgh, Pittsburgh, PA 15260

The web-based virtual mass spectrometry laboratory (VMSL) was developed as a cross-disciplinary tool to help students learn mass spectrometry and problem solving. This free resource is accessible to a wide range of students and provides an active learning opportunity to the student through life-like case studies. The user is guided to solutions to these problems through an inquiry approach that allows them to control virtual mass spectrometers that return real data that they need to manipulate, analysze and interpret.

Introduction

Over the last two decades, mass spectrometry (MS) has evolved into a powerful cross-disciplinary analytical technique. As a result, many students and scientists seek to learn more about MS and how one operates such an instrument for chemical analysis. Unfortunately, mass spectrometers are expensive to purchase and to operate and they are not found in every laboratory. In addition, mass spectrometrists are often not available for instruction and even when they are, only one student at a time can effectively learn how to operate the mass spectrometer by hands-on operation. The Virtual Mass Spectrometry Laboratory (VMSL) (*1*) was developed to remove these restraints so that many undergraduate students in the classroom or at home can simultaneously learn the basics of how to use mass spectrometers to solve problems. Because the resource is web-based and free, this active learning exercise occurs at the student's convenience.

The VMSL is an interactive, internet educational tool that teaches mass spectrometry by letting students solve real problems from different disciplines without actually going to a MS lab. The VMSL allows each student to choose and operate several virtual mass spectrometers, to acquire many mass spectra and to manipulate and interpret the data collected. We generated case studies in the areas of proteomics, polymer characterization and small molecule analysis. In each of our studies, our goal is to achieve a real-to-life educational experience. In this chapter, we explore several case studies including: Protein Identification by both MS and MS/MS, Polymer Analysis, Caffeine Analysis, General Anesthetics and Cocaine in Hair. We end with an evaluation and our conclusions of the VMSL as a teaching tool.

Some undergraduate students who have used the VMSL have remarked that they would have preferred using a real instrument. We respond by saying, "Where would you go to have such an experience"? The VMSL is indeed not the replacement for ideal instruction involving hands-on operation of a mass spectrometer and training by a real person, but given the lack of these resources, it is the next best active learning tool. We have tried to make the VMSL program a process that is close to reality. We want students to read about the MS techniques and instruments before entering the lab, but we also want students to actively pursue a solution to a problem by study and explicit control of the virtual mass spectrometers.

How the Site Works

The VMSL is a teaching resource designed to allow students to operate simulated mass spectrometers, including deciding what data to collect and under which type of instrument settings. The "raw data" is provided to the student who must then manipulate it in appropriate ways in order to collect evidence to answer the question posed in the case study. All aspects of the resource are available to any student with access to the internet via a standard web browser interface (it has been optimized for use with Microsoft's Internet Explorer).

The raw data files, which can be both, large in size and numerous in scope, remain at all times on the resource's servers, while the information passed to the client is in the form of small text-based files or gif images. This strategy was consciously implemented in order to keep bandwidth requirements to a minimum and to minimize the resource's dependencies on nuanced differences in hardware architecture of each client machine. By using server-side routines to interpret client requests and to parse the data accordingly, it becomes trivial for a client to have the most recent version of the resource – it's the one posted on the web. The downside of not having a downloadable version that could be run independent of access to the net is increasingly less important as net access continues to expand its reach into all nooks and crannies of our society.

While details of how the resource is coded are beyond the scope of this document, the method employed (Figure 1) is for the client to issue a request to the server to conduct a "certain experiment". The server routine extracts the appropriate data file from the database server, formats the data as necessary according to client directives, and then sends a text and/or image file for display on the client's screen. The GC-MS case studies use custom-written algorithms for extracting, manipulating, and formatting data stored in a generic form (using custom software to recast proprietary formatted data into more easily manipulated data for a database) while the MALDI-TOF and LC-MS experiments rely on the instrument manufacturer's software package (from the instrument on which the data was collected) to manipulate and format data prior to sending the small text/gif files to the client. As the astute reader will have ascertained, all possible experiments that a client might request will have been actually run on that instrument and those raw data file results parked in the database server for future use. In other words, the VMSL resource uses real data and does not simulate any of the data manipulated by the client.

Educational Philosophy

The resource was designed and created with several educational objectives in mind. First, we knew that faculty had access to mass spectral data in hard copy form but that most students had no or very limited opportunities to operate a mass spectrometer. Furthermore, those students who did have hands on experience usually were limited to GC-MS instruments only, and often with limited access and little opportunity to "play around". Second, we were intrigued with case-based learning, (*2*) the added-interest such brings to students and the use of mass spectrometry to solve scientific questions that the public at large could relate to. Lastly, from our extensive histories in the use of mass spectrometers to provide data to answer specific questions, we knew that multiple paths can be followed to collect appropriate data for any one question and a single scripted approach was not indicative of real life.

The approach we adopted utilized several strategies. For example, each mass spectrometry "challenge" was framed in a life-like problem-based scenario that was designed to both motivate students to seek the answer while providing them with a rich learning environment. We also desired to create a virtual environment that mimicked, as near as we could practically achieve, the actual

Figure 1. Schematic diagram showing the initial storage of many MS files onto the server during development of a case study and the later use of the data as transferred from the server to the user during the operation of the VMSL.(Adapted with permission from ICTE 2002: International Conference on New Technologies in Education. Copyright 2002.)

environment a scientist would experience. Therefore, in addition to the background reading and description of the problem, there is an instrument room (Figure 2) where the student selects which mass spectrometer she will use, a "wet-lab" area where decisions about sample preparation must be enacted, and the instrument control area where the mass spectrometer operational parameters must be selected. Finally there is a data manipulation portion where acquired raw mass spectral data files can be analyzed in various ways in order to both determine and document the answer to the question posed in the case-study.

The approach to a problem is not scripted and each student can follow their own path, including the amount of background or preparative work done, choosing the consensus approach (to sample preparation and instrument parameter settings) based on their understanding of the background reading or exploring for themselves what the different combinations of sample preparation and/or parameter setting provides in terms of quality of data. Because there is an infinite number of pathways that could be followed, and realizing that all those experiments would have to have already been run and archived on the database, some student selections for options will return comments such as "That [option] is a reasonable choice but the corresponding data is not available; please make another selection.". Thus the site was designed and implemented with an inquiry approach very much in mind.

Protein Identification

Protein identification by MS is probably occurring thousands of times each day throughout the world. The VMSL Protein Identification case study is based on the powerful research tool of identifying a protein based on a peptide map and/or mass spectrometry/mass spectrometry (MS/MS) data. To solve the VMSL Protein Identification case study, the operator must determine what animal species donated a sample of serum albumin extracted from blood. The student must first choose from a set of several unknown blood samples from which the serum albumin protein will be extracted. Next, the student is given a choice of what instrument and what method will be used to solve the problem. In the example that follows, a matrix-assisted laser desorption ionization (MALDI) time-of-flight (TOF) mass spectrometer is chosen from the instrument room which features four instruments (Figure 2). Alternatively the student could pick the liquid chromatography (LC) electrospray ionization (ESI) mass spectrometer in the bottom left hand corner of Figure 2 since we also incorporated this method into the VMSL. The MALDI TOF MS method creates a peptide map to identify the protein while the LC ESI MS method uses MS/MS fragmentation data, but given the limited space, it is not discuss herein.

After choosing the mass spectrometer the student must prepare the calibrant, matrix and protein for the MS analysis. The proteins are reduced, alkylated and digested with trypsin using proper reaction conditions to form unique peptide fragments. Aliquots of the calibrant/matrix mixture and the protein digest/matrix mixture are placed onto a MALDI plate and allowed to dry.

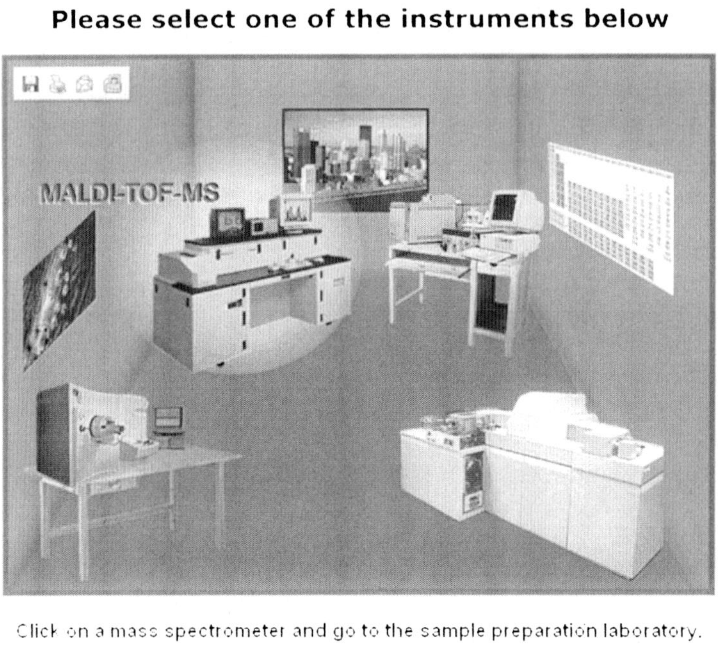

Figure 2. The VMSL instrument laboratory.

To collect a quality mass spectrum, the student must optimize the mass spectrometer in the data *Acquisition Mode*. As shown in Figure 3 poor acquisition parameters result in a "bad" mass spectrum. That is, the mass spectrum lacks a useful signal. Note that the *Acquisition Mode* user interface allows for a variety of inputs such as well position, laser type, laser power, delay time, laser shots, mass range and more. At this point, the operator must further tune these parameters, acquire a new spectrum and then calibrate the mass spectrometer. Results from the optimization are displayed in the *Data Explore Mode* shown in Figure 4 for a six peptide calibrant. The *Data Explore Mode* allows the student to manipulate the data and examine the quality of the spectrum. For example, the student may "zoom" in and examine individual mass peaks if so desired and the VMSL server will return a new GIF image of this portion of the spectrum. In this process, the student may initially collect many examples of poor quality mass spectra in the tuning process, but only the collection of a quality mass spectrum will allow the student to obtain a good calibration so that she can proceed to the analysis of real samples.

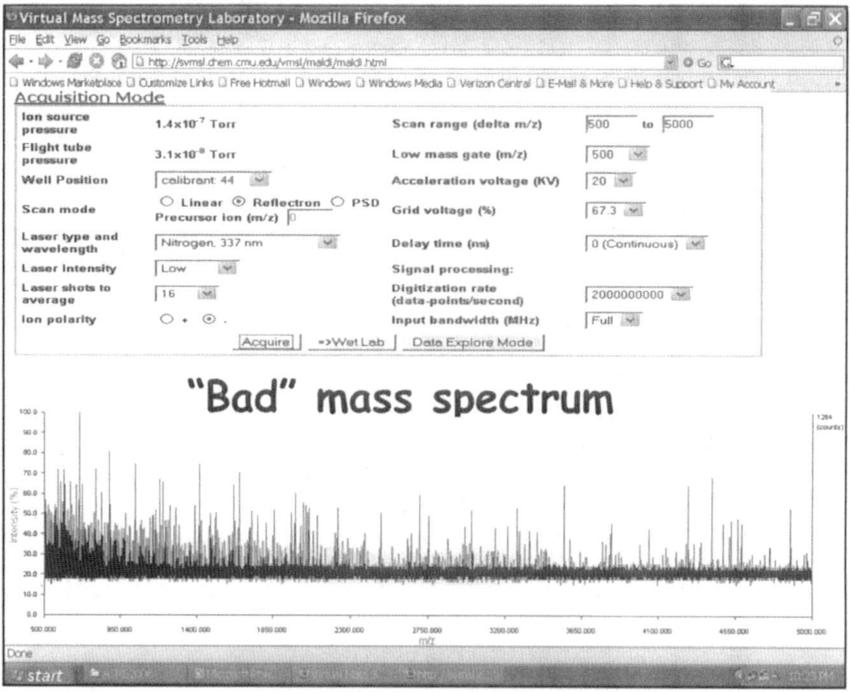

Figure 3. A MALDI TOF mass spectrum that was not optimized. The peptide signal is not distinguishable from noise.

Once the calibration is saved, the student can proceed to the identification of the serum albumin sample. A MALDI mass spectrum collected from one of the unknown protein digests creates a peptide mass map which can be used to identify the animal which donated the protein with a high probability. Since each species has a unique amino acid sequence making up their serum albumin, the set of peptide products formed from the digestion will be unique for each animal species. Accurately determining the masses of these peptides will allow the student to identify the protein as serum albumin and thus specific animal species. We include about nine species in this VMSL study and more can be added.

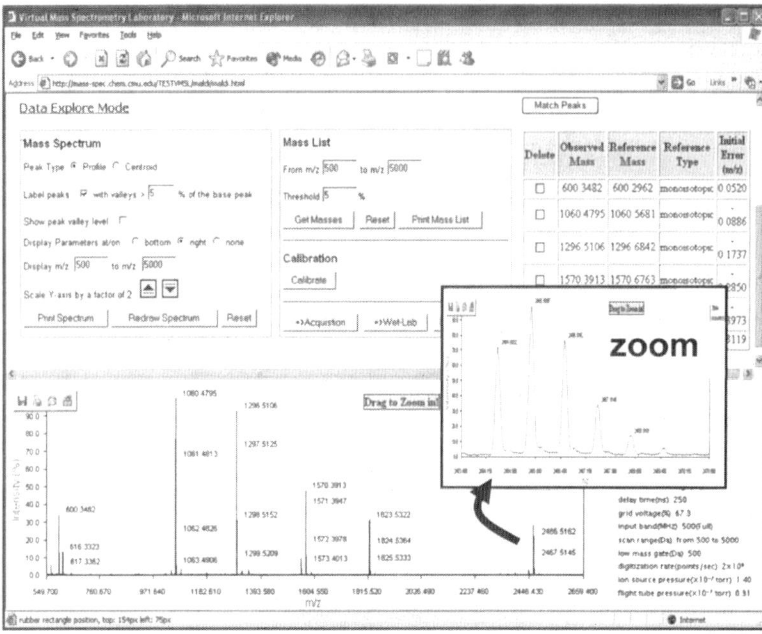

Figure 4. MALDI TOF mass spectrum of a six peptide calibrant. The zoom feature shows the spectrum quality under high resolution. The upper right panel shows the calibration table.

The final step requires the student to submit a *mass list* of peptides to a protein database search engine. The user can choose from several search engines such as *MS-Fit* from the University of California at San Francisco (*3*) or *Protein Mass Fingerprint* in MASCOT (*4*), a commercial search engine. Both are free and on the internet as referenced. The final hit of the correct species brings a significant feeling of the positive reinforcement that comes about from solving the case study.

Polymer Analysis

Two problems are listed on the VMSL site for the Polymer Analysis:

Problem one: Lance Legstrong wants to enter this year's Tour de France with an advanced bicycle tire called SuperRoll™. Unfortunately, the recent manufacturing batch of these tires failed. The key to the enhance performance is the use of a polystyrene additive at 0.02% of the formulation. The polystyrene additive increases the Tg (Glass transition temperature), but it must have a polydispersity of 1.03 ± 0.05 and a Mn of 4700 ± 100. Determine what is wrong with additive #1 used in the last batch and determine which polymer additive meets the specifications required for the SuperRoll tires.

Problem two: Calculate the polydispersity index (PDI) for the different polymer samples sets of polyethylene glycol (PEG) and poly-methyl-methacrylate (PMMA) provided and determine the monomer unit and end groups.

To solve problem number one, the student must determine Mn, Mw, Mz and the polydispersity index (PDI)(*5*) of several polymer additives to find out why the tires failed and what additive should have been used in the correct tire formulation. The student must first select an additive from the five possibilities and then proceed to the *Acquisition Mode* to tune the mass spectrometer and collect the mass spectrum. From the spectrum, the student must then determine whether the polymer tested is the correct additive. Figure 5 shows A MALDI TOF mass spectrum of polystyrene showing the proper additive with acceptable Mn, Mw, Mz and PDI to be used in the tire formulation. Note that to the right side of the spectrum a list of the parameters used to collect the data is given so that the student may print out this data and turn it in to the instructor or keep it for a record.

To solve problem two, the student must select a polymer sample from one of two polymer types: polyethylene glycol (PEG) and poly-methyl-methacrylate (PMMA). The MALDI TOF mass spectrometer is chosen from the MS room shown in Figure 2. Note that a GC-MS could not be used to solve this case study due to volatility and a message will appear if such is chosen, to tell the student to make another selection. The user must next prepare the polymer sample and calibrant and an appropriate matrix using the correct solvent, concentration, volume and then proceed to the data *Acquisition Mode.* In the *Acquisition Mode* the instrument must be tuned to achieve an optimized mass spectrum.

Note the user interface allows for a variety of inputs (well position, laser type, laser power, laser shots, mass range and more) as in the other instrument control panels and in some cases a movie frame allows the user to explore the laboratory and to view the mass spectrometer. After acquisition of a polymer spectrum the student goes to the *Data Explore Mode* to manipulate the data and calculate the Mw, Mn, Mz and the polydispersity (PDI) for the polymer sample. The student might collect many examples of poor quality mass spectra, but only the collection of quality mass spectra will allow the student to solve the problem. The monomer unit can be determine by calculating the difference between polymers of different chain lengths and the end groups determined by calculating the expected MW for various end groups.

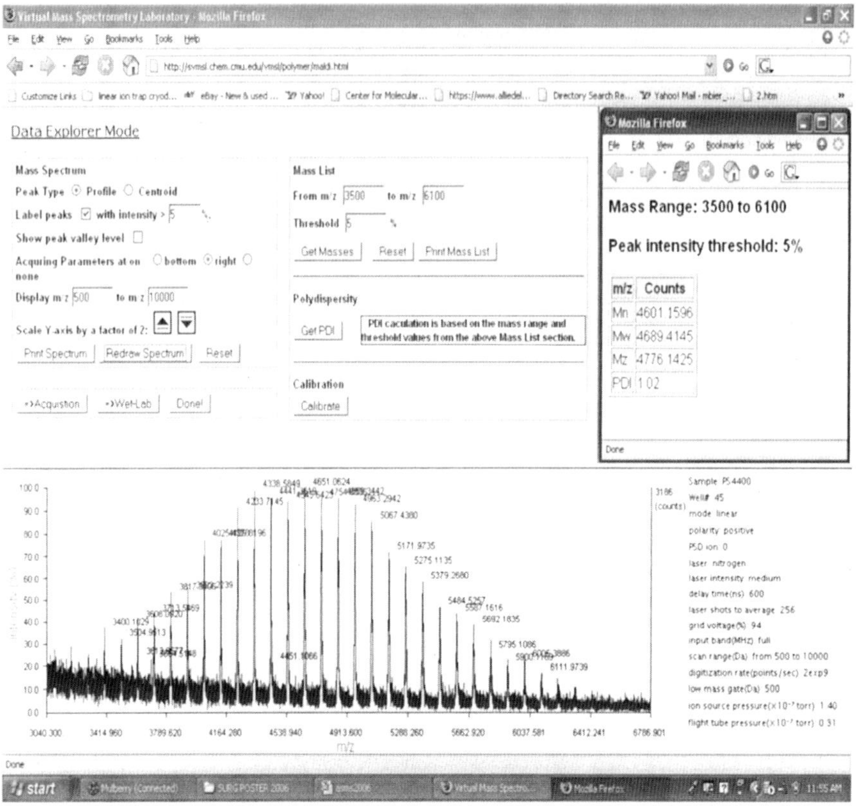

Figure 5. A MALDI TOF mass spectrum of polystyrene showing the proper additive with acceptable Mn, Mw, Mz and PDI to be used in the tire formulation.

General Anesthetics

The *General Anesthetics* case study is summarized on the web site as follows:

> A historian from the National Museum of Civil War Medicine in Frederick, Maryland, discovers what he believes to be a mid-1800's medical kit on a Civil War battlefield near Knoxville in eastern Tennessee. Inside the kit is a small vial of liquid whose identifying label can no longer be read due to faded ink. Suspecting that the vial contains an anesthetic used in the Civil War, he sends the liquid to a laboratory to be examined. What is in the vial?

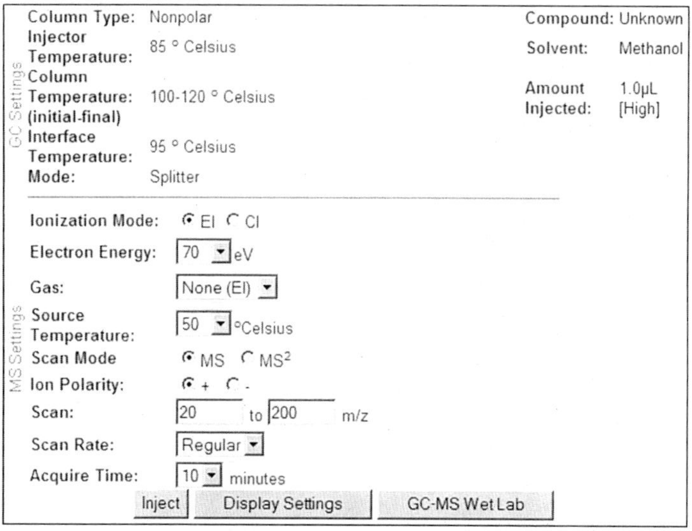

Figure 6. The virtual GC-MS instrument control screen

Included in the optional background reading is a brief history of general anesthetics, focused on the time up to and during the American Civil War as well as more modern general anesthetics. A student who chooses to read these short background sections will know what the general anesthetic of choice was during the Civil War and therefore a hypothesis of the compound that is in the vial. Importantly, the one-page synopsis of the case study also contains a link to a two-page *Student Manual* that delineates the questions that a student should answer as they work their way through the case study and gives them guidance about specific issues to focus on in both designing and interpreting their experiments. From these materials, the students will know that they need to select the GC-MS from the *Instrument Room* in order to most efficiently answer this case study. After preparing the first sample in the wet lab the student encounters the instrument control screen shown in Figure 6. In order to minimize the complexity, the GC settings are fixed and only representative MS settings are user adjustable. After "injecting" the sample, the data is returned which can be manipulated in any number of ways (Figure 7). At this point the user can zoom in on a portion of the TIC, examine the mass spectrum that corresponds to any point in the TIC, or return to the wet lab to prepare another sample for injection and further data collection.

We now routinely use this case study as a homework assignment in second semester organic chemistry, at the point were mass spectrometry is introduced.

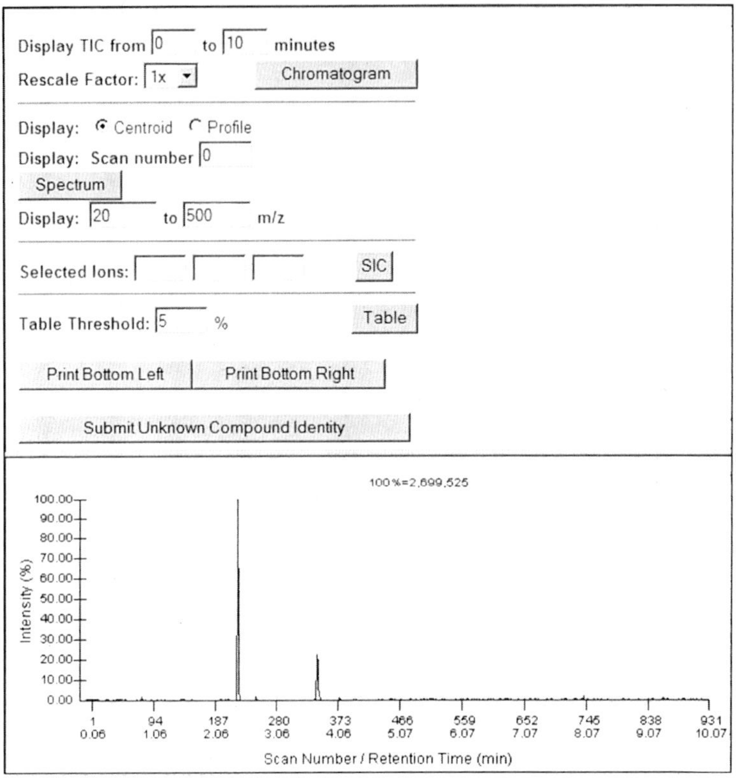

Figure 7. GC-MS data manipulation area.

The case study is readily solved by the sophomores in the course, where they are intrigued by the ability to collect their own data, manipulate it and reach a conclusion. The students prepare a lab report summarizing what they did and providing retention time and mass spectral documentation to support the identification they made. By using the text and gif formats built into the resource, the students are easily able to incorporate the data into a MS-Word lab report that they submit to the instructor electronically. Averaging the student opinion surveys that were completed each of the past two years, 88% of the students who were assigned the General Anesthetic case study as a required homework indicated that they would like to solve another case using this method of instruction. To aid the instructor, additional resources available include a Teacher's Manual as well as pre- and post-quizzes to gauge learning gains. The caffeine case study (discussed next) was created to aid in the introduction of the VMSL and GC-MS to an organic lecture class.

Caffeine Study

When we first started using the General Anesthetic case study for sophomores, it became apparent that an instructor needed to explain a variety of aspects of GC-MS to the students. Whereas they may have already understood that mass spectrometers "weigh" ions, they had little appreciation of how a GC worked, how a MS worked, or how the two worked in tandem. In addition, a GC-MS experiment returns an immense amount of raw data which the students were unfamiliar in terms of knowing how to zero in on the relevant data.

In order to not use the General Anesthetic case study (so that it could be used as a take-home quiz) to introduce the students to GC-MS instrument operation, types of data returned, and data manipulation, we found it advantageous to create an additional case-study for that purpose. Therefore we created a GC-MS case-study on the caffeine extracted from coffee; this extraction is a common organic laboratory course experiment. A caffeine VMSL experiment might then serve multiple purposes – useful for the introduction of VMSL and GC-MS for certain classes, and as an add-on exercise for laboratory sections already performing the caffeine extraction.

For second semester organic chemistry, we now routinely use the caffeine case-study as part of our two lecture days dedicated to the use of mass spectrometry towards structure proof. The first day introduces many of the fundamental ideas of mass spectrometry (converting molecules into ions and fragments, weighing and counting them, data presentation and interpretation). Included in this discussion are sample purification methods and the coupling of a purification technique to the mass spectrometer. The second day we actually run the virtual GC-MS as a group, with students choosing what samples to run and which instrumental parameters to change. Invariably, we briefly discuss choices with the discussion dictated by the students' curiosity and the options they have.

During the class room demonstration of the VMSL via the caffeine case study, we discuss instrumental choices, the necessity for consistency, the tools available for data manipulation, and what kind of data is available. The "three dimensional" data available from GC-MS is always a challenge to get across and so this case-study has an explicit section dealing with that issue with care taken to define TIC and SIC (total and selected ion chromatograms respectively). During these discussions, retention time and fragmentation patterns that are signatures of individual molecules are emphasized. The discussion always concludes with explicit mention of documenting conclusions and of how to do that, of the utility of authentic spectra, and the differences between spectra on your instrument versus that from a library. The students are always amazed and frustrated that real-life chemistry is not perfectly black and white and well-behaved, with infinite values for signal-to-noise ratios. The ability to "play" with real data during a lecture is a valuable outcome of this in-class activity.

Cocaine in Hair

The third and most complex GC-MS based case-study in the VMSL is *Cocaine in Hair*. It is not uncommon for a young person seeking their first employment with a major company to be asked to submit a sample of hair for analysis. As explained on the VMSL web site:

> Hair Analysis is relatively new technique to the forensic world. Hair analysis can sometimes provide a long term chronological record of a person's selected drug use. This is because drugs like cocaine (or marijuana, phenylcyclidine (PCP), amphetamines, among others) are incorporated into the hair matrix at the time of ingestion and remain in the hair until it is cut or otherwise seriously damaged. Urine and blood tests are the standard and accepted methods of drug analysis today, but they are only useful if the sample is taken within 2-3 days of ingestion. Hair analysis can show from months to years of drug usage depending upon the length of the subject's hair. It is common today, even for summer-time jobs in the corporate world, for employers to require a hair sample from any job applicant. These hair samples are being tested for the presence of drugs of abuse.

This case study is set-up in a fashion parallel to the *General Anesthetics* one discussed above. Background reading includes sections on "Cocaine – A Brief Overview", "A History of the Study of Cocaine in Hair", "The Controversy of Hair Testing", "General Procedures for Hair Testing", "Detailed Procedure to Extract Compounds", and "A Few Chemical Structures". Sample preparation and instrument parameter selection are identical to the earlier case studies on this instrument.

Due to the number of compounds extracted, a very "busy" set of data can be obtained, and each hair sample will be extremely different from other samples of hair. (To work with hair containing cocaine, we soaked hair randomly collected from a barbershop in a cocaine solution, rinsed and dried it and then treated both the spike and un-spiked samples by the same extraction procedure). The rich data set returned requires the user be comfortable with SICs and to have done the requisite background work to select the correct fragment ions to search for cocaine signatures. In other words, the two previously discussed case studies should be completed before this one is attempted.

The implementation of this case study is still being improved. We would like to improve both the spiking of the hair samples with cocaine, and the procedure to extract it. We would like to add additional samples to increase the number of data sets (even though any one user will still see only one option for the "unknown"). Finally, we need to improve the background material on the electron ionization mass spectrum of cocaine in addition to providing library spectra and access to "authentic" samples for the user to run in parallel with their experiment.

Evaluation

The evaluation of the VMSL Internet tool has been a difficult task for several reasons. First a good method of measuring the effectiveness of the VMSL in teaching must be developed and for good statistics hundreds to thousands of students are required in the evaluation pool. For typical class sizes in which the VMSL was evaluated there have been less than thirty students. In addition, the method used in the evaluation was not exactly the same for each class. This was not possible because the same instructors were not used for each evaluation. At times, the VMSL was used at institutions external to Carnegie Mellon and the University of Pittsburgh so the surroundings and the instructors were not the same. Furthermore, even after a request for consistency, for evaluation requirements, the instructors often varied from the teacher's guide or verbal instruction and made modifications to the student's VMSL experience. This was understandable in most cases as the teachers made changes due to the student's background, whether that is a discipline background in chemistry verses biology or a difference in educational experiences between sophomores or juniors or seniors. In addition, classroom periods were different and the lesson had to compensate for that factor as well. In the end, we evaluated the VMSL using two completely different methods. Method one was a direct survey and request for feedback comments. Method two required that the student take a *Pre-quiz* before solving the cases study and once they finished, for example the *Protein Identification* case study, they took a *Post-quiz.* The *Pre-quiz* was identical to the *Post-quiz,* but the students were not told this until the end. Again, evaluation anomalies arose from this test because students would take the *Pre-quiz* several times before proceeding to the case study and these results had to be omitted. Different instructor's prepared their students in different way for this latter evaluation. Some instructor made the quizzes count towards their grade, other made it extra-credit, while still others made the quizzes just for fun. Ideally, the quizzes were to count towards the students' grade so that they would have a reason to learn from using the VMSL laboratory.

The results from evaluation method one revealed that many students did not prepare well for the VMSL and treated the lesson more like a typical recipe driven laboratory rather than a real-life experiment. This is somewhat expected since almost none of the students had any sort of real-life lab experience, but, of course, a real scientist must do some upfront reading of the literature before attempting to identify a protein for the first time. We also learned that there is an activation energy barrier for both students and teacher to use the VMSL. Once over the barrier people were generally happy, but like anything new, the experience was not necessarily straight forward as one would like. As expected, good students will ask questions and not proceed without a good level of understanding while poor students will not ask questions and instead try to proceed by trial and error. Users often expected a tutorial with all the answers and were not accustomed to a teaching tool that mimics real-life. One student wrote. "Provide a bibliography as I have never done a protein digestion! ". We thought this to be reasonable suggestion, although one might question why the student did not look up a reference or two for themselves.

The results from methodology two, after elimination of many pre- and post-quiz entries due to inaccuracies in testing, was that there was indeed an improvement in the twenty question protein identification quiz results after the VMSL teaching experience, but it was not dramatic. For a sample size of 126 students, they received an average % correct ± sigma of 56.2 ± 17.4 % for the *Pre-Quiz*. For a sample size of 49 students, they received an average % correct ± sigma of 69.8 ± 18.0 % for the *Post-Quiz*. This was an improvement in score of 13.6% equal to two more correct answers.

Conclusions

We would like to conclude by noting some of the important things we have learned from our involvement in the creation and deployment of the VMSL. Designing, programming, refining, and implementing the VMSL has been an incredibly complex, demanding, and rewarding exercise, much more so than we anticipated. The resource has interfaces for several mass spectrometers and the coding has been designed so that additional data sets/case studies can be added with little modification of the instrumental parameter selection and data manipulation screens. While adding case studies is possible, considerable effort is needed to prepare the background material and to run the actual experiments that provide the mass spectra for the database.

The creation of a resource is what we set out to do and what we accomplished. However we little realized that the "marketing" of such a resource, even as novel and as simple for a user to implement as the VMSL, is a time-intensive and long-term process that really began as the grant was ending. Even so mundane a task as keeping the server software and hardware up-to-date and reliably accessible was not fully appreciated at the outset. New users often have questions, and good suggestions for modest improvement, and advances in knowledge and web-accessible resources all converge to demonstrate the need for on-going resources, even if just for a maintenance-level effort. One of our ever-present challenges is how to continue to promote and maintain this resource. The NDSL and other "electronic aggregators" are a fantastic resource to promote links to resources. Whereas as a traditional library maintains a book once it is deposited therein, is there a similar resource that would maintain an electronic project such as the VMSL (and not just maintain a link to the project)?

With well-established syllabi and commercially published books, it is perhaps not surprising that encouraging adoption of an electronic resource such as the VMSL for inclusion in their courses by additional faculty is a challenge.

We find that many faculty are interested when they learn about it, but that promoting the resource takes time and resources that extend beyond what a typical NSF grant can provide. Efforts like the ASDL project, and electronic, living textbooks such as that under development for Physical Chemistry are two approaches that can expand the reach of local developed resources such as the VMSL. However, it is likely that even more effort would be beneficial in this area.

Another active learning approach for mass spectrometry teaching is the direct implementation of remote-control of mass spectrometers from the classroom. Bier implemented such an approach in 1999 and has continued to use this teaching method and the VMSL in his *Mass Spectrometry* class at Carnegie Mellon. The remote control methodology has also been used in the *STaRBURSTT Cyber-Instrumentation Consortium (6)* for X-ray crystallography and a number of local consortia have done so for NMR. There are pros and cons to such approaches over a virtual resource such as the VMSL, and a best case scenario may indeed be to employ both approaches. The field of active learning of experimental science using web-enabling approaches will undoubtedly continue to evolve over the coming decade.

Acknowledgements

The VMSL project was funded by NSF DUE-9950867, Carnegie Mellon University and the University of Pittsburgh. Additional funding for the mass spectrometry instrumentation at Carnegie Mellon University came from NSF CHE-9808188 and NSF DBI-9729351. Current support of the VMSL project is from USDA 2005-38422-15925.

References

1. Virtual Mass Spectrometry Laboratory. http://svmsl.chem.cmu.edu (accessed Dec 6, 2006)
2. National Center for Case Study Teraching in Science. http://ublib.buffalo.edu/libraries/projects/cases/case.html (accessed Dec 6, 2006)
3. ProteinProspector. http://prospector.ucsf.edu (accessed Dec 6, 2006)
4. Matrix Science. http://www.matrixscience.com (accessed Dec 6, 2006)
5. William Reusch. Virtual Textbook of Organic Chemistry. http://www.cem.msu.edu/~reusch/VirtualText/polymers.htm (accessed Dec 6, 2006)
6. The STaRBURSTT CyberInstrumentation Consortium. http://www.starburstt.org (accessed Dec 6, 2006)

Chapter 14

The Analytical Sciences Digital Library: A Useful Resource for Active Learning

Cynthia K. Larive[1] and Theodore Kuwana[2]

[1]Department of Chemistry, University of California, Riverside, CA 92521
[2]Department of Chemistry, University of Kansas, Lawrence, KS 66045

The Analytical Sciences Digital Library (ASDL) is a collection of peer-reviewed, web-based resources, related to chemical measurements and instrumentation. This website is freely accessible at www.asdlib.org. ASDL is an important supporting resource for those interested in adopting problem-based or other discovery learning curricular resources. Materials in the ASDL collection include animations, on-line texts, PowerPoint™ lecture slides, tutorials, simulations, virtual experiments and examples of educational best practices. Sites are categorized to allow for easy browsing in the areas of Class Materials, Virtual Labs, Techniques, Applications and Teaching Resources. The collection can also be searched using a keyword search function. ASDL also publishes Online Articles in the ares of e-Courseware, e-Labware, e-Educational Practices, and e-Undergraduate Research Highlights.

What is ASDL?

The Analytical Sciences Digital Library (ASDL) has been supported by grants from the National Science Foundation (NSF) Division of Undergraduate Education (DUE) and the NSF Directorate of Mathematics and Physical Sciences (MPS). The collection went online in November 2002 as www.asdlib.org. Its development and operation were described in an A-page article in the journal Analytical Chemistry in June 2004 (*1*). ASDL is one of a number of digital library collections contained in the National Science Foundation National Science Digital Library (NSDL) program. NSDL is the Nation's online library for education and research in Science, Technology, Engineering, and Mathematics (STEM). NSDL was created to provide access to high quality resources and tools that support teaching and learning in STEM areas of science. Although NSDL supports educational activities at all levels (K-16 and beyond), ASDL is primarily focused on content for university-level undergraduate and graduate students. It fills a unique niche within the NSDL by supporting educators, students and practitioners of the analytical sciences.

A goal of ASDL is to facilitate active learning in chemical measurements and instrumentation by providing links to peer-reviewed online web-based resources to the analytical sciences community. The ASDL collection (asdlib) grew out of a series of NSF-sponsored workshops on "Curricular Developments in the Analytical Sciences" organized by Ted Kuwana in the late 1990's. A major recommendation of the workshop report was that the analytical sciences academic community "develop context-based curricula that incorporate problem-based learning (PBL)" and other discovery learning approaches (*2*). For many faculty, the hurdle for adoption of discovery learning practices was identified as the lack of supporting resources, best practices and other curricular materials that could be adapted by interested faculty as appropriate for their local classroom setting (*3*). ASDL was established in part to satisfy this need. The ASDL website, www.asdlib.org, went online November 2002 and continues to evolve to serve the needs of the analytical science education community.

What are the contents of the Asdlib collection?

The ASDL portal is illustrated in Figure 1. Asdlib is a collection of peer-reviewed, web-based resources, related to chemical measurements and instrumentation. It is a freely accessible website at www.asdlib.org. Because the resources in the main asdlib collection are already in the public domain, copyright is not an issue. All of the materials in the ASDL collection are peer reviewed by experts in the field and are annotated to provide a short description of the site content and the level of student for which the site is most appropriate.

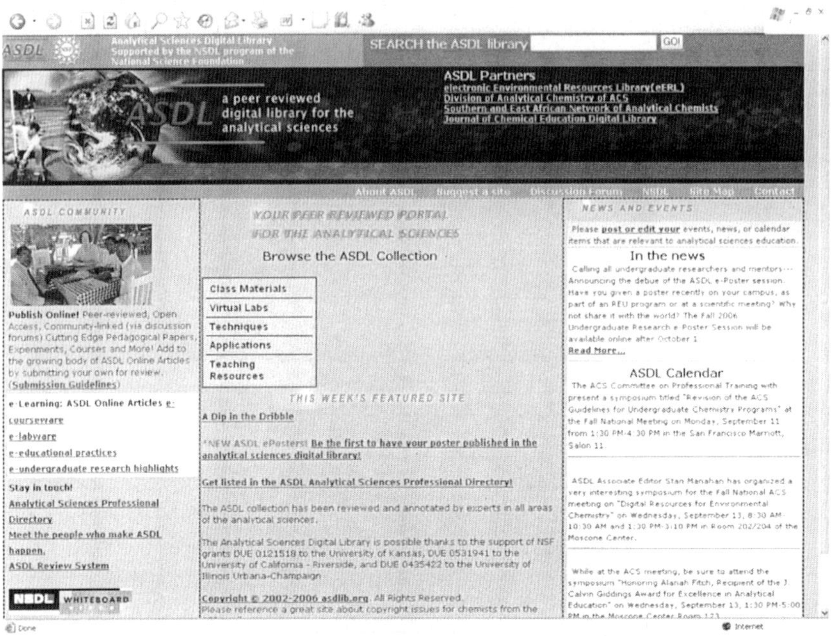

Figure 1. A snapshot of the ASDL portal.
(See page 1 of color inserts.)

Asdlib content includes links to more than 325 annotated websites. Materials in the main collection, accessed by the drop-down menus in the center of the portal, include animations, on-line texts, PowerPoint™ lecture slides, tutorials, simulations and virtual experiments. Sites are categorized to allow for easy browsing in the areas of Class Materials, Virtual Labs, Techniques, Applications and Teaching Resources. The collection can also be searched using a keyword search function.

In addition to providing annotated links to web-based content, ASDL has recently begun publication of Online Articles. ASDL publishes online articles in the areas of e-Courseware, e-Labware, e-Educational Practices, and e-Undergraduate Research Highlights. The online articles are peer-reviewed articles published in the online section of ASDL. Under the Creative Commons license [http://creativecommons.org/license/], the author(s) can retain copyright ownership or release the material to the Public Domain. In the former case, the author(s) specify the conditions such as "you are free to copy, distribute, display and perform the work," including opton of making derivative works. In the latter case, once a work is dedicated for released to the Public Domain, the author(s) has relinquished all rights to ownership and control of the material in question. ASDL recommends author(s), as self-designated, retain copyright ownership and give permission for ASDL to place the material, as appropriate, in the online section.

Electronic access allows publication formats not supported by traditional print journals. For example, asdlib contains many computer animations that can help students visualize complicated processes. Electronic access also enables access to virtual experiments or instruments, either in preparation for a hands on lab or to supplement the training of students who do not have access to high end instrumentation, for example mass spectrometers (*4*). An interactive discussion board is associated with each article enabling community-based discussion of pedagogical approaches related to the article, highlighting the advantage of online publishing. The UnderGraduate Research articles provides an opportunity for undergraduate authors to publish a summary of their work, and since they can retain copyright, should not preclude subsequent publication of their work in traditional research journals. We expect that this feature will not only provide an opportunity for dissemination of research results, but it will also serve as a mechanism to establish a community of undergraduate analytical science researchers and the faculty who serve as their mentors. ASDL also provides an opportunity for analytical science practitioners in instrustry and government labs to share their experiences by developing case studies for undergraduate analytical science courses.

How does ASDL operate?

ASDL editorial functions are implemented using a journal model in which AEs identify potential new web content, submit that website for peer review, evaluate the reviews, make a decision about whether the site merits inclusion in ASDL, and write an annotation describing the site. While suggestion of sites by community members are welcome, most of the sites in the asdlib collection have been identified by AEs through Internet searches. The editorial staff has found the Highlight Concept Hierarchy Development search engine, http://highlight.njit.edu/, to be a particularly effective tool for identifying new sites for consideration for inclusion in ASDL. The software for this search engine was developed at New Jersey Institute of Technology by a group

Table I. The ASDL Editorial Structure

Cynthia Larive	Editor-in-chief	Magnetic resonance Mass spectrometry Bioanalytical	University of California-Riverside
Ted Kuwana	Managing Director	eUnderGraduate Research FIA & biosensors	University of Kansas (emeritus)
Heather Bullen	Associate Editor	eUnderGraduate Research ePosters	University of Western Kentucky
Alanah Fitch	Associate Editor	Environmental applications Remote access instrumentation	Loyola University – Chicago
Richard Kelly	Associate Editor	eUnderGraduate Research ePosters	East Stroudsberg State University
Carol Korzeniewski	Associate Editor	Electrochemistry Surface methods	Texas Tech University
Patricia Mabrouk	Associate Editor	Education innovations Nanomaterials applications	Northeastern University
Stanley Manahan	Associate Editor	Environmental applications	University of Missouri (emeritus)
Steven Petrovik	Associate Editor	Quantitative analysis Food & beverages applications	Southern Oregon University
Richard Safferstein	Associate Editor	Forensics applications	Widener University
Alexander Scheeline	Editor for On-line Articles	e-Courseware, e-Labware, e-Educational practices	University of Illinois Urbana-Champaign
Mark Vitha	Associate Editor	Separations	Drake University

including Yi-fang Brook Wu and Michael Bieber with support of the NSDL program.

Once a resource has been reviewed and annotated, it is accepted into the ASDL collection. The editorial and review functions are carried out electronically using a password locked web-based utility accessed from the ASDL portal. ASDL functions thanks to a cadre of Associate Editors (AEs) who generously donate their time to build the ASDL collection. ASDL AEs are faculty with expertise spanning a wide range of topics related to chemical measurements and instrumentation and an interest in innovatve educational methods. The current ASDL AE team is listed in Table 1. One of the significant challenges that ASDL faces is the transient nature of Internet materials, either due to content revision with a new URL, or deletion of the website by the source from the World Wide Web (WWW). If the loss rate is greater than additions, the collection will not remain viable. One solution is to archive in a separate location the essential information of sites so that they can be tracked back to their sources. When they are no longer available on the web, inquiry to the source could be made to ask whether ASDL can either download with permission into ASDL/asdlib, or to ask for revisions that can be posted and accessed again on the Web. The best solution is of course to identify and add new peer-reviewed sites to the collection at a rate above the deficit.

ASDL Community Features

Several ASDL features have been developed to support the user community including, News, Events and the ASDL Professional Directory. The News column highlights news relevant to ASDL users such as student fellowship and travel award opportunites. The Events section summarizes upcoming items of interest, for example symposia related to analytical science education at scienfic conferences. A unique community feature of ASDL is the Analytical Sciences Professional Directory. This feature was started as a way for faculty, especially those at predominately undergraduate institutions, to find colleagues with similar research and teaching interests. At the request of practicing analytical scientists we recently opened the directory to those in industrial and government positions. We are somewhat surprised at the popularity of this feature and are in the process of implementing browsing by categories (faculty, industrial, government) and adding search capability within the directory. Because the directory information includes database entries for contact information, we have a mechanism for contacting participating scientists about new ASDL initiatives. In addition, the information they provide about their research interests and expertise allows ASDL AEs to access these ASDL users as content reviewers.

How can ASDL content support active learning?

There are several ways in which the materials in the ASDL collection can support the use of active learning in the analytical sciences curriculum. One mechanism is by providing an archive of resource materials specially targeted for instructors. In ASDL these resources are found primarily in the Teaching Resources section of the main collection and in the Online Articles section. The Active Learning section of the Teaching Resources category contains many items that discuss strategies for successfully incorporating active learning into lecture and laboratory courses. For purposes of illustration, a few of these items are described in the following section. The web-paper by Donald Paulson and Jennifer Faust presents of survey of active learning techniques that can be integrated into college-level lecture courses (*5*). Paulson, a chemist, wrote this so that the suggestions offered are sufficiently generic to be applicable to any field. Exercises for students to perform individually or in groups are summarized along with strategies designed to help students provide immediate feedback to the instructor about their level of understanding of difficult concepts and stimulate critical thinking. Another ASDL resource provides a tutorial for using PBL in large classes. Examples of the use of small group, self-directed, self-assessed PBL in tutorless groups in the chemical engineering program at McMaster University are presented (*6*).

For many faculty, a barrier to implementing problem-based and other active learning approaches is in developing appropriate assessment tools. A resource in the ASDL archive that addresses this topic is the paper " Student Assessment in Problem-based Learning" by Jeffrey Nowak and Jonathan Plucker. In their paper Nowak and Plucker note that the lack of attention to assessment is a weakness in many applications of PBL (*7*). They discuss the importance of aligning the assessment with the instructional activities and providing students with reasonable guidelines about the instructor's expectations.

Several of the ASDL Online Articles also provide useful information for support of active learning. At the time of this writing, the e-Labware category contains three examples of online publications that support problem-based learning in the analytical sciences laboratory. A laboratory manual for forensic science, developed by Rob Thompson, contains a list of 10 different forensic-based chemical analysis experiments with detailed information for students and instructors (*8*). Along a similar vein, Julie Stenken has developed a laboratory experiment for the instrumental analysis laboratory involving the authentication of paintings (*9*). Lisa Holland's article on capillary electrophoresis describes how to assemble an inexpensive, yet reliable capillary electrophoresis apparatus and provides experiments to demonstrate electrodynamically-driven separations

(*10*). An e-Educational Practices article by Alex Scheeline describes his experiences implementing active learning in his introductory analytical chemistry course for graduate students (*11*). An important feature of each on-line article is a discussion forum that allows members of the community to post improvements and discuss other issues related to the article content.

One of the most important ways in which ASDL can support active learning is by providing reliable resources for students engaged in problem solving. Active learning approaches engage students and often involve them in developing their own chemical analysis methods. Asdlib contains links to several applications notes for sample preparation and analysis. In addition, links are provided to several databases including the NIST webbook which Chemical and physical property data on over 30,000 compounds and The SDBS spectral database for organic compounds produced by the Japanese National Institute of Advanced Industrial Science and Technology (12,13). The SDBS database contains ^{1}H and ^{13}C NMR spectra (14,000 and 12,300, respectively) ESI mass spectra (22,900), FT-IR (49,800), laser Raman (3500) and ESR (2500) spectra for 32,000 compounds.

Partnerships and self-sufficiency

From ASDL's perspective, there are two types of partnerships. One is a recipricol linking with an organization's website, via its URL, to complement each others content. An example is the Journal of Chemical Education's digital library that collects and catalogs digital web resources in areas of chemistry education [http://chemed.chem..wisc.edu/JCEDLib/index.html]. Another is the electronic Environmental Resources Library [http://www.eerl.org/] that contains link to online material of environmental and sustainability resources for community college educators and students, and practitioners in the field. There is little, if any, overlap in content between ASDL and these partners.

Under the National Science Digital Library (NSDL) program as funded by the Division of Undergraduate Research (DUE) of NSF, the dictum is that collections such as ASDL must become self-sufficient at the end of the current grant period. Thus, another type of partnership being sought by ASDL is with an organization that has financial resources to fund us at a sustainable level. Because the essential infrastructure for ASDL, including the site software and operational developments are now in place, the cost is primarily to support a webmaster, continued marketing and meeting of principals. As of this writing, negotations are underway with a professional organization to establish such a partnership in which ASDL will serve also as a platform to enhance web contents for both groups.

The global perspective of ASDL

In November 2005, we began using the free web tracking service provided by Google called Google Analytics. Google Analytics counts visits, rather than hits, eliminating false counts generated by web crawlers. This service also provides information about how users come to the site, which pages are accessed most frequently and how users navigate the site. As shown in Table 2, roughly 2000 unique visitors access www.asdlib.org each month, viewing approximately 3 pages per visit.

Table 2. ASDL Use Statistics

Month		*Vists*	*Page Views*
December	2005	1690	4181
January	2006	1950	5324
February	2006	1862	5073
March	2006	2052	6184
April	2006	2158	6140
May	2006	1874	5233
June	2006	1449	4198
July	2006	1332	3892
August	2006	1890	6129

The Google Analytics service also provides information about the location of ASDL users in graphical form, as illustrated in Figure 2. In this example for the month of May, the circles on the map reveal the location of users while the size of the circle is proportional to the number of users accessing www.asdlib.org from that location. While we expected to find a large concentration of users in the United States, we have been surprised by the world-wide access of ASDL and its content. These maps illustrate both the global nature of the Internet and the world-wide interest in analytical science, a subject that is especially important to developing countries.

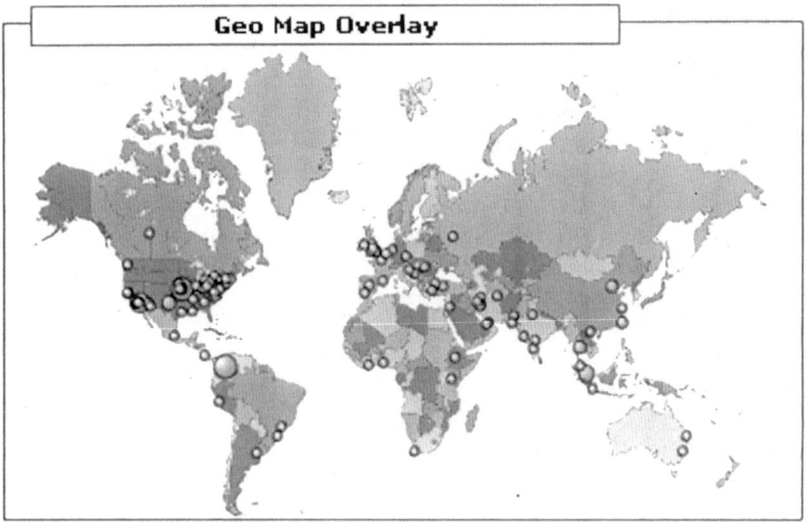

Figure 2. Google Analytics results for May 2006 (See page 1 of color inserts.)

References

1. Larive, C.; Chalk, S.; Kuwana, T. *Anal. Chem.* **2004**, *76*, 398A.
2. Kuwana, T. Curricular Developments in the Analytical Sciences. http://www.chem.ku.edu/TKuwana/CurricularDevelopment/ (accessed Aug. 22, 2006).
3. Wenzel, T. J. *Anal. Chem.* **1998**, *70*, 790A.
4. Bier, M.; Grabowski, J. Virtual Mass Spectrometry Laboratory. http://mass-spec.chem.cmu.edu/VMSL/ (accessed Sept. 5, 2006).
5. Paulson, D. L.; Faust, J. L. Active Learning for the College Classroom. http://www.calstatela.edu/dept/chem/chem2/Active/ (accessed Aug. 23, 2006).
6. Woods, D. R. Problem-based Learning, Especially in the Context of Large Classes. http://www.chemeng.mcmaster.ca/pbl/pbl.htm (accessed Aug. 23, 2006).
7. Nowak, J. A.; Plucker, J. A. Student Assessment in Problem-based Learning. http://www.indiana.edu/~legobots/q515/pbl.html (accessed Aug. 23, 2006).
8. Thompson. R. Forensic Science Laboratory; Analytical Sciences Digital Library: www.asdlib.org, #10002, Oct. 04, 2004.

9. Stenken, J. A. Adapting the Chemical Analysis of Paintings; Analytical Sciences Digital Library: www.asdlib.org, # 10038, Feb. 16, 2006.
10. Holland. L. A. Microseparations Distance CE; Analytical Sciences Digital Library: www.asdlib.org, #10031, Oct. 04, 2005.
11. Scheeline, A. Active Learning in Advanced Analytical Chemistry, a first year course for graduate students; Analytical Sciences Digital Library: www.asdlib.org, #10039, July 22, 2005.
12. NIST Webbook. http://webbook.nist.gov/ (accessed Sept. 06, 2006).
13. Spectral Database for Organic Compounds, SDBS. http://www.aist.go.jp/RIODB/SDBS/cgi-bin/direct_frame_top.cgi?lang=eng (accessed Sept. 06,2006).

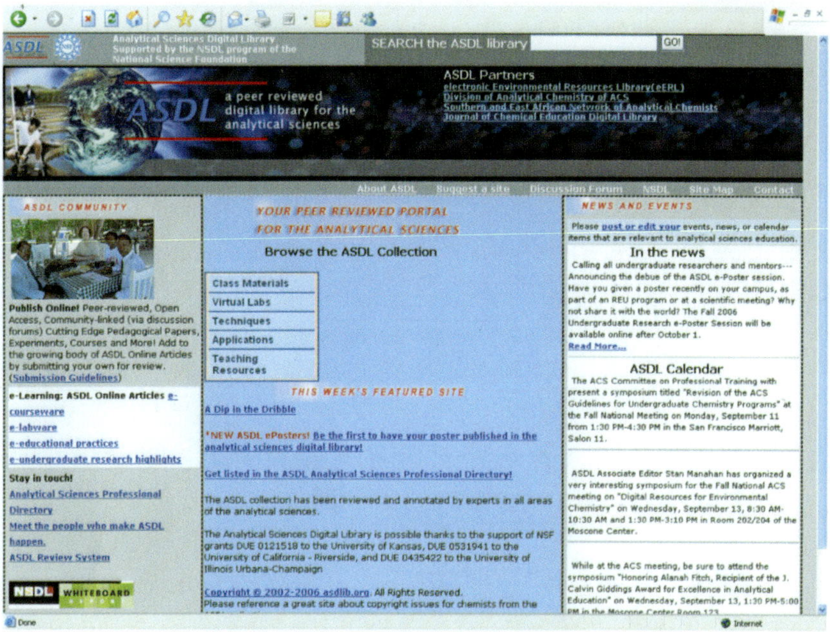

Figure 14.1. A snapshot of the ASDL portal.

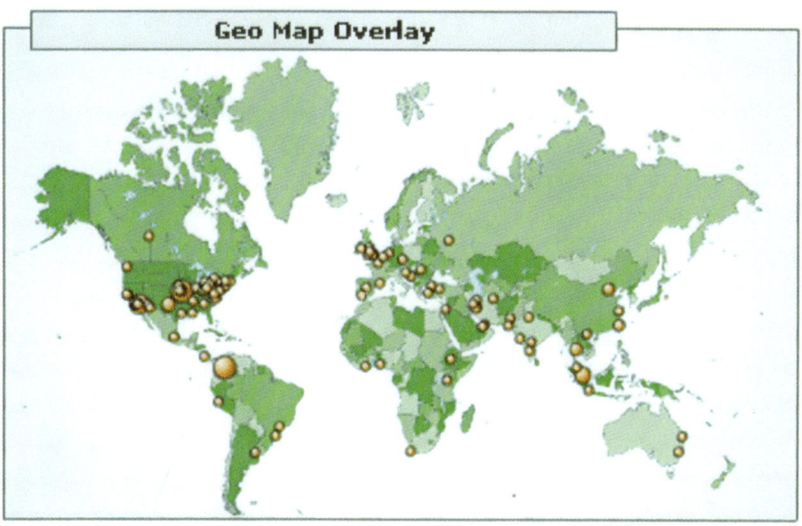

Figure 14.2. Google Analytics results for May 2006

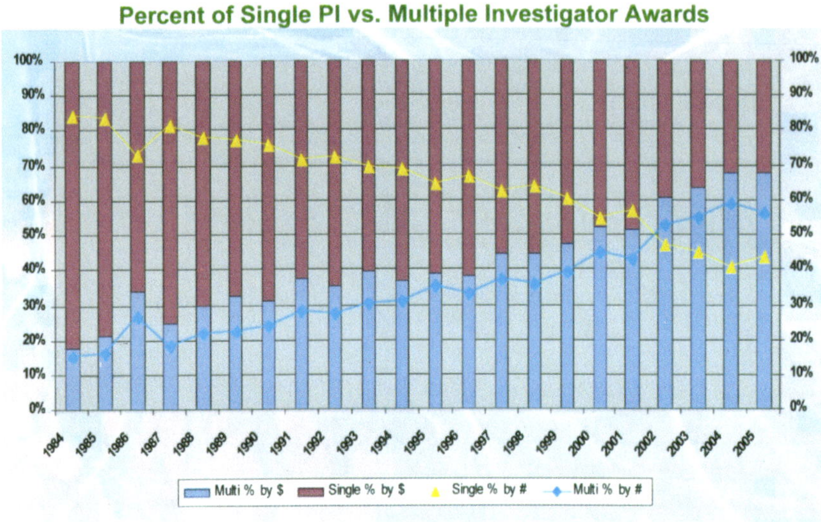

Figure 19.1. Trends in single investigator vs. multiple investigator awards. (Adapted with permission from reference 2. Copyright 2006.)

Figure 19.2. Cartoon depicting elements of a successful collaboration. (Adapted with permission from reference 4. Copyright 2006.)

Chapter 15

Establishing the Foundations of Analytical Chemistry Using a Web-Based Format

R. K. Gilpin and C. S. Gilpin

Department of Chemistry, Wright State University, 3640 Colonel Glenn Highway, Dayton, OH 45323

This chapter discusses the development and operation of an online course that is currently used to teach beginning analytical principles including conversions, data collection and handling, simple principles of acid-base, solubility, and redox equilibria, and beginning spectrometric measurements. The authors review some of the areas that can/do cause difficulty for first time online instructors, which they encountered during the development and operation of the course and how these problems have been solved.

Introduction

Over the last decade, web-based instruction has gone from a curiosity to an accepted and often highly productive teaching tool. Today, there are many successful examples of totally and partially web-based courses that are taught in both the distance learning and the on-campus modes at small, intermediate, and large universities throughout the Nation. Unfortunately, when the natural sciences are involved, the acceptance and use of web-based teaching has not been met with open arms, but more often with passive resistance to open hostility. Although there are a variety of reasons given by both teachers and students why they do not like online activities, a number of them are either philosophical or operational based *(1,2)*.

Major concerns of many chemists, physicist, and allied professional educators are related to either their belief that: 1) teaching quantitative thinking and experimental measurement skills cannot be reproduced in a web-based format, but need personal contact and hands-on experiences, or 2) scientific observations are dynamic in that they are often distorted by the presence of systematic and random errors. Likewise, the major concerns of students often are related to the feeling that: 1) they are not well prepared in terms of mathematics, especially when it involves selecting the appropriate scientific relationship/mathematical formulation, and 2) they do not see direct links between the mathematics, the chemical or physical concepts, and their past, present, or future lives.

In order to address chemical educator concerns about online teaching, it is important to understand why they arise. In the case of quantitative thinking, the expectation is that a student can proceed from a set of experimental observations and results to a quantitative description of the events formulated in a manner that is consistent with the first principles that describe or govern them. Secondly, individual results can change from measurement to measurement and they are complicated further by the possibility of both systematic and random errors as well as non-ideal effects. Thus, the student must not only learn the underlying concept that theoretically describes the observation, but they also must be cognizant of the fact that it contains some level of uncertainty by virtue of making the measurement and that there are effects that are not easily describable by first principle relationships *(1,2)*.

In the case of many students, they often are ill prepared in terms of their mathematical skills or the math that they were taught in high school was not done in a conceptual framework that demonstrated its relationship to solving real problems. The result of poor preparation and the disconnect between mathematical theory and its subsequent application, is that a majority of students are uncomfortable solving descriptive-based problems that provide the necessary information, but are not formulated in the final mathematical solution (i.e., the chemical or physical relationship) *(1,3,4)*.

The focus of this chapter is to discuss the design and operation of an online first term chemistry course. Key elements in designing useful and challenging courses are organization, maintenance, and innovation. Like other facets of our lives, effective communication, which involves the bidirectional flow of ideas, is vital to both teacher and student success in the online setting. It is important that communication is not confused with course delivery, which is typically the unidirectional flow of information. The posting of reading materials and lecture notes are examples of course delivery, whereas learning materials that are interactive, dynamic, and provide immediate feedback are communication tools. Some useful examples of the latter are self-testing modules, on-demand help, and other supplemental learning resources such as dynamic pictorials, interactive graphics, and computer simulations that emulate basic laboratory operations and measurements *(3,5-7)*. The successful use of these and other innovative web-based tools have been discussed by many scholars in a variety of fields including the physical sciences, engineering, and medicine *(1-14)*. A common thought is that, when used properly, the internet is a dynamic learning resource that provides new opportunities to introduce ideas and concepts in a more graphical and interactive fashion, which enhances the learning experience not detract from it *(15)*.

Course Structure and Organization

The overall structure and organization of web-based courses are extremely important as is the design of the web site and pages that are used to navigate and retrieve information. Students have widely varying levels of computer skills and a well designed web-course must meet the needs of all students irrespective of their computer literacy *(16)*. These might seem like obvious statements that would not be challenged by many. Nevertheless, in practice they are often not followed. However, when dynamic web-based tools are designed and used properly they have a stimulating influence on students *(8,12,17)*.

Web Page Design

In terms of general layout, consistency and repetition are good elements as is simplicity. The initial web-page (i.e., the home navigation page) should not be filled with clutter and it should be user friendly. An interesting example that illustrates this idea is to compare the ease of obtaining information using the Web-page of the internet search engine, Google™ vs. most university and company homepages. It is important to keep in mind that elaborate web-page designs in combination with very detailed descriptive materials often do more harm than good in terms of immediately "turning-off" students. It is important that course designers consider the learners' needs not their own preferences *(18-20)*. There are a myriad of reasons why students enroll in online courses *(21)*, but none of them are to struggle with overly ambitious web-design. It is a good "rule-of-thumb" to assume that most beginning college students are less familiar than they think in terms of the operation of their computer's hardware and software.

A second consideration is related to how students are initially introduced to the course website verses their repeated usage of the site. The first few visits should not overwhelm them in intricacy or content bulk. It is often better to introduce students to the course website slowly in terms of the total number of items that they first see (i.e., assuming this option is available with the delivery platform being used). This is accomplished by turning-on or posting only a few items when the course first begins. The initial items might include a course syllabus, rules and policies of the course, the first one or two sets of lecture notes, one or two interactive exercises and a complete listing of the scheduled exams. Also, a contact number and e-mail address should be two of the more distinctive items that appear on the course homepage. Likewise, it is a good idea to have these on assignment, activity, and other resource pages as well. One reason many students give when asked why they fear web-based courses is the belief that they will be abandoned when problems occur *(15,22)*. After a few weeks, when the students have become comfortable navigating and using the site materials, the total course content can be turned-on. An alternative approach to the staggered introduction of material is to design the course in terms of smaller modules of learning activities and to turn these on and off as the students proceed sequentially through the course *(23)*.

Both staggered and sequential approaches work well once students have become familiar with the course web-site design and navigational features. However, the sequential approach works best when students have a conceptual ideal of how each of the individual parts fit into the overall scheduling of important course activities such as exams and other assignment due dates *(20)*. Posting these on the course's homepage from the start is appreciated by most students and minimizes misunderstandings in terms of grading expectations.

In addition to keeping the homepage simple but informative, the overall design of the web-site should be user-friendly in terms of navigating between pages. Although many students are experienced at retrieving information from the Web, a large number are not. A well designed internet course must be designed to meet the diverse needs of students so that its does not overwhelm those with minimal computer skills and turnoff ones with higher skills *(16)*. Self-guiding prompts that are distinctive (i.e., bolded, underlined, a different color, etc.) and concise (e.g., "get the notes," "help," "return to homepage," etc.) are useful navigating tools that meet the needs of most. Likewise, when designing a course web-site, it is important to keep in mind that most students fit into one of two categories. They either have difficulty navigating through several layers of web-pages or they do not wish to contend with overly elaborate "page-nesting" when they are studying *(10,13,24)*.

Course Structure

Although there are several different philosophies when considering the overall structure of the beginning chemistry course sequence in terms of content organization, this topic is beyond the scope of the current chapter. Nevertheless, the basic approach used by the authors is to teach the first term course using the underlying principles that are the common underpinnings of analytical chemistry, including conversions, data collection and handling, simple principles of acid-base, solubility, and redox equilibria, and beginning spectrometric measurements. In the addition, the authors have taught this course in both the web-intensive and totally online modes and have found that a majority of students prefer the web-intensive mode based on end-of-term surveys as well as in terms of the numbers of students that enroll in each. As such, the current chapter emphasizes useful approaches and techniques for structuring/organizing and delivering materials via the web-intensive mode that have been employed by the authors in a number of beginning classes they have taught for several years with a combined enrollment of more than 2000 students. Although several different approaches have been tried in terms of assignments and testing (i.e., the amount of weight placed on online activities vs. traditional in-class exams), in all cases there have been high linear correlations between the students' final grades in the class and the amount of time they have spent using the various types of online learning resources and in carrying out interactive activities and simulations. This is based on a detailed analysis of the delivery platform's daily student logs and course grading records *(1,3,6)*.

One of the important and unifying features of the content delivery system has been the development and use of learning modules. Breaking (i.e., grouping) the course into a few subunits (i.e., referred to throughout the remainder of the chapter as learning modules) has been a useful technique of controlling information flow and guiding student progress. For most quarter to semester courses, three to six major groupings are about right in number. Fewer groupings typically do not provide enough flexibility in organizing the course content and more groupings complicate information flow and consistency as well as overwhelm many beginning students *(19,23)*. In the first term chemistry course taught by the authors, three major groupings are used and in the second term course, four provide a better fit to the overall content being taught.

Shown in Figure 1 is a flowchart that summarizes the various elements of a first term, ACS accredited, chemistry course organized into three major units of activities that each cover two to three chapters from the textbook that is used currently in the course. Each learning module is designed around a database of questions (i.e., typically 30 to 50) that are related to fundamental concepts from which the computer randomly selects ten. Each question is mathematically formulated where it can have an unlimited number of quantitative solutions. Because the computer randomly generates values for the variables within a question, each time that it is shown, it is unique. Likewise, since the database of questions is relatively large, questions are randomly selected, and new variables appear each time it is presented. Therefore, students cannot simply memorize an answer, but they must learn the basic approach behind each question in order to be successful when they are given an in-class exam on the material *(25)*. Although there are other possible ways of organizing a first term chemistry course besides question-based, this approach has the advantage of helping

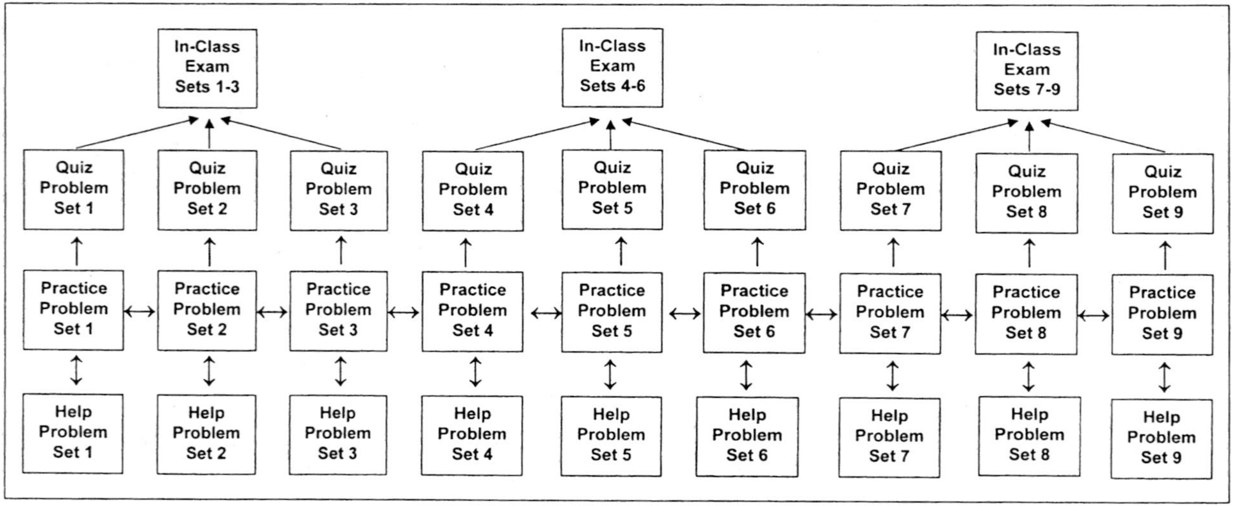

Figure 1. Flowchart illustrating the overall design of an interactive beginning chemistry course based on nine learning sets divided into three learning modules.

beginning students overcome one of their greatest fears of chemistry, their inability to solve problems given a descriptive set of information. Likewise, problem solving promotes the development of cognitive skills that go beyond the chemistry they are studying. This latter aspect is important since most students taking beginning chemistry are from other fields in the natural and physical sciences as well as engineering and medicine.

As noted previously, the first term chemistry course is organized into three learning modules each containing three question sets. Because the course being discussed is taught in the Web enhanced mode not the totally online mode a significant portion of the students' final grades are based on traditional in-class exams. Once the students have completed a group of three question sets (i.e., a learning module), they meet in a traditional classroom setting to take a computer generated exam that is made up of randomly selected questions from the three sets of questions, (e.g., an in-class test containing thirty questions is made up of ten questions selected randomly by the computer from each of the individual sets). For the approximately 1600 students who have taken the first term course, there is almost a linear correlation between the amount of time they have spent on the online exercises (i.e., data obtained from the detailed computer logs) and the grade they earn on the in-class exams, as well as their final course grade. Likewise, similar levels of success have been observed in terms of the second term chemistry course that contains four learning modules each of which is based on two interactive questions sets and other interactive experimental simulation activities. Additional details concerning these aspects of the course are discussed in later sections of the chapter. The second term course, unlike the first term course, which teaches basic analytical chemistry concepts, covers simple organic and physical chemistry concepts. Since both courses are organized similarly (i.e., based around interactive question sets and experimental simulations) and to avoid any confusion between them, the remainder of the current chapter will use only the first term course to illustrate important details and features of the both.

The three learning modules shown in Figure 1 are organized in increasing difficulty. The intent of the first module is: 1) to review fundamental concepts that beginning students should have been covered in pre-college chemistry, and 2) to provide students with an opportunity to become comfortable using the online web-delivery platform, the organization and use of the learning modules and accompanying online resources, and the formulation and submission of responses. It is important to realize that many students enter a college level beginning chemistry class with highly varying levels of skills in not only their pre-college chemistry, but also in their mathematics and physics training. Likewise, students have remarkably diverse backgrounds in terms of the general use of computer hardware, operating platforms, and data manipulating and word processing software. As such, the course must not only bring students to a common level in terms of their pre-college chemistry preparation but in allied areas of importance *(7,16,20)*.

Summarized in Table 1 are representative questions randomly selected from the nine problem sets. Two questions were taken from each of the three major groupings. In the initial set (Set 1 in Figure 1), the questions are based on simple concepts and mathematical manipulations. Working through this problem set provides students the opportunity to become familiar with the organization of the course's website, as well as the online materials and their use to study fundamental concepts. As students proceed through the course, the difficulty level of the questions increases in the problem sets. In some instances, information is included in the problem that may not be needed to work it. By doing this, the students must develop a higher level of understanding about the particular principle or relationship that is needed to solve the problem.

Interactive Help

After reviewing the online course material for a given section, the students begin with the practice set questions, where they enter their solutions and submit them for grading. The computer grades the submitted answer sheet and returns it as an interactive screen where the student can review their answers; decide whether they need help with a particular question; and obtain an example of the correct procedure for solving the missed question. This is done by using an interactive on-screen "Help" link. An example help screen is shown Figure 2 for a randomly selected question taken from the measurement and conversion problem set in learning module 1 (i.e., the initial question set shown in Figure 1).

It is important to note that during the development and early evolution stages of the first term chemistry course (i.e., the first two times that it was offered), the practice problem sets did not contain the interactive "Help" feature. As a result of this missing feature, from classes with enrollments of about 160-180 students, several dozen of them would request help each week during scheduled office hours and many more would request help via e-mail questions. With the addition of the "Help" feature and other interactive learning tools, very few students now find it necessary to seek help that is directly related to working the online chemistry problems. However, they still seek help related to system, computer, and other operational problems. This latter aspect, which is covered in detail later in the chapter, is perhaps the most underestimated aspect of successfully maintaining a web-based course.

When the students are using the practice sets, they may work and submit randomly generated questions as many times and as often as they wish. Once the students feel that they have mastered the material in a problem set, they then proceed to the corresponding online quiz for that unit. Unlike the practice sets, the quizzes have a time limit and may be taken only once. The grade is recorded in a spreadsheet that is always available to the student, for review, throughout the course. The students overall course grade is made-up of online quizzes, other

Table 1. Representative questions taken from each of the three major groups of learning modules.

Group	Question
1	How many centimeters are there in 69.93 feet?
1	You are carrying out a scientific measurement that requires you to record the laboratory's temperature. Unfortunately, the only thermometer that you have available is calibrated in Fahrenheit and you need to report it in Kelvin. As such, the lab temperature is 68.4 °F, what is this value in K?
2	How many milliliters of a 0.1 M solution of NaOH are required to carry out a titration of 2.922 grams of a polyprotic acid that has a formula weight of 172.0 and exactly reach its second end point?
2	How many milligrams of Ag^+ are present in 172 mL of water containing a precipitant of silver chloride that has a K_{sp} of 1.8 x 10^{-10}?
3	$H_2C_2O_4 + MnO_4^- \rightarrow CO_2 + Mn^{2+}$ The above reaction takes place in an acidic solution. After balancing the reaction, how many moles of H^+ are needed for each mole of MnO_4^- that reacts?
3	If the outer electron configuration of iron is $4s^23d^6$, when the Fe^{2+} ion is formed its electron configuration is?

interactive activities, and in-class traditional exams. Although the quizzes are given on an honor basis, students that do not do the online materials for themselves using this system, do poorly on the in-class exams. Since the online quizzes count only about 40% of the overall grade, it is not possible to use these by themselves to pass the course.

Other Interactive Materials

In addition to the problem sets and quizzes discussed above, several other types of learning resources are used in the course. These range from simple text files of lecture notes and other instructional aids, to dynamic pictorials, Excel-based simulations, and fully executable standalone programs. The latter two of these, the Excel-based simulations and the fully executable Windows-based programs, represent the most recent additions to the course materials in its third and fourth development stages of evolution. An example of one of the dynamic

Example Question:
How many meters are there in 76.46 yards?

Answer:
You need to remember the following conversion factors.

$1\ in = 2.54\ cm$
$1\ m = 100\ cm$
$1\ ft = 12\ in$
$1\ yd = 3\ ft$

Using these conversion factors, you can write the following equation:

$$76.46yd \times \frac{3ft}{1yd} \times \frac{12in}{1ft} \times \frac{2.54cm}{1in} \times \frac{1m}{100cm} = 69.92m$$

Performing the mathematics and canceling identical units in the numerator and denominator gives a result of 69.92 m. This is one way to find the answer. You can use different conversion factors if they are easier for you to remember.

Figure 2. Example of an interactive help screen for the measurement and conversion problems set contained in learning module 1.

pictorials used in the early stages of the course is an interactive Periodic Table. By clicking on an element, information about the element and its physical properties is displayed in fields that appear just below the Table. Many similar interactive pictorials and animations have been, and are being used daily by students throughout the Nation to explore/illustrate a host of scientific and engineering principles. Since many useful examples of these can be obtained online using one of the reliable Internet search engines, additional details and discussion of them are beyond the scope of the current chapter.

In addition to various interactive pictorials and animations, both Excel-based simulations and fully executable standalone Windows-based programs have been developed and used by the authors in their online courses. These custom software packages are designed to emulate various classical and modern instrumental laboratory methods. The most important and unique features of their construction are that they 1) are self-generating in terms of unique parameters and solutions, 2) contain systematic and random errors, and 3) they incorporate nonlinear effects commonly observed in carrying out the actual laboratory experiment. To date, a variety of programs have been developed for emulating classical gravimetric and volumetric methods, extractions, chromatographic separations, and infrared and mass spectrometric measurements *(1,3,6)*. Not all of these are used at the introductory chemistry level, but they span a range of skill levels from the beginning first and second term chemistry courses to more advanced graduate level analytical courses taught by the authors. Currently, ten simulations have been employed in the beginning first and second term chemistry courses. Some of them are supplemental laboratory exercises and others are related to data processing and manipulation.

Shown in Figure 3 are screen views that a student would see as they progress through one of the Excel-based simulation, which is a simple volumetric titration experiment. At the start, the liquid in the beaker below the burette is colorless and after the endpoint has been reached it is deep pink. Between these two points, the intensity of color change and its dispersion through the solution vary in the same fashion as observed in an actual titration. This emulation is designed to simulate the color changes one would see when carrying out the titration in the laboratory using phenolphthalein as the acid-base indicator. This experiment has been and is being used to introduce students to the basic concepts of volumetric analysis prior to carrying out the experiment in the laboratory setting. To date, several thousand on-screen practice titrations have been performed by students over the last few years. Currently, a study is in progress to evaluate the effectiveness of on-screen practice simulations on student performance in terms of the quality of their data and time needed for them to complete the actual titration.

Maintaining Online Courses

The discipline and content area of online material and how it is designed has a significant affect on the amount of effort needed to keep the course running smoothly. First-time online educators often underestimate the amount of day-to-day work required to keep the class functioning seamlessly, especially as it relates to maintaining immediate to near immediate feedback (*7*). Likewise, first time students often underestimate the amount of effort that they must put forth on a regular basis to be successful in online learning. It is not unusual for university educators to simply reduce their previous in-class lecture notes, accompanying reference materials, and homework sets to files that are posted on the Web with expectations that this will significantly improve their class, increase student performance, and make life easier for both themselves and the students. Unfortunately, they often find that the number of day-to-day questions received from students are an order of magnitude greater and the student level of dissatisfaction significantly increased *(10,20,21)*.

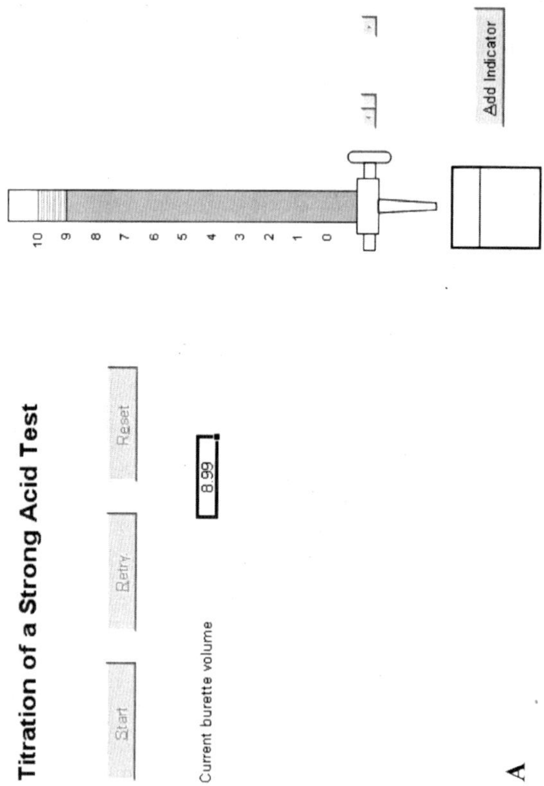
Titration of a Strong Acid Test
Start
Retry
Reset
Current burette volume
8.99
10
9
8
7
6
5
4
3
2
1
0
Add Indicator

A

B

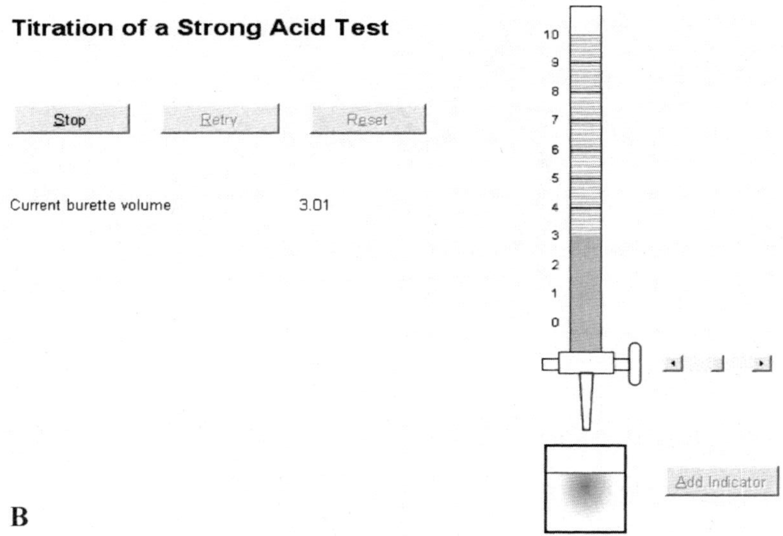

Figure 3. Emulation of a strong-acid base titration.
Continued on next page.

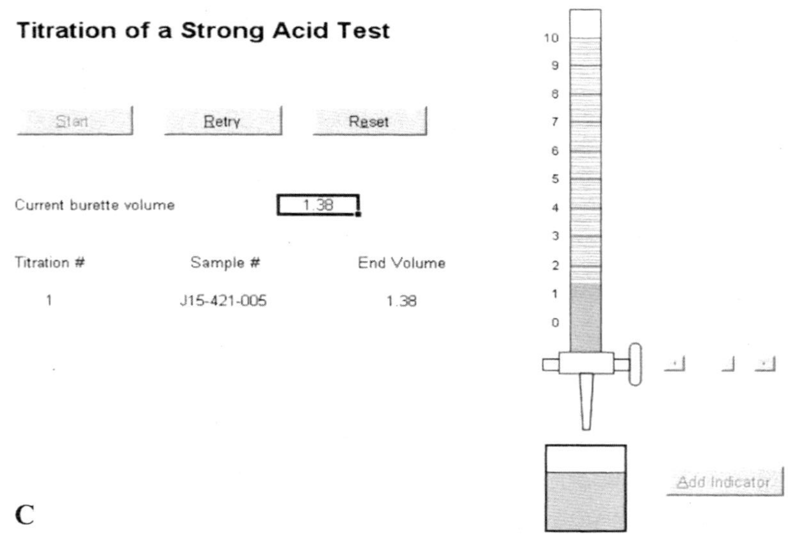

Figure 3. Continued.

Operational Problems

Within an online course, there are several areas that require very minimal amounts of oversight and there are others that are extremely intensive in terms of their day-to-day demands. If the course's content material has been developed properly with appropriate learning resources, evaluation and testing resources that are dynamic and self-grading, and an interactive feedback system to provide immediate help with the solutions to incorrectly submitted responses, the day-to-day operation of the course requires little to no personal attention from the instructor or assistant to deal with content related questions. Developing this type of course requires much more work prior to posting it online, but the rewards to both the educator and students are large. Unfortunately, the design and operation of this type of course places greater demands on the reliability of operating systems (i.e., the College or University delivery platform and the end-user's computer and operating system), as well as available Internet bandwidth. Also it is more susceptible to software differences, updates, and system modifications. Thus, if the course is properly designed, most of the day-to-day student questions are operational/system based not discipline/content based.

Many of the operational problems result from: 1) constant updates/upgrades that university's do to maintain their web-delivery platforms, 2) imposed security protocols to protect against invasive programs (e.g., viruses and spyware), 3) variability in the operating systems used by the students, and 4) students' background and skills in PC-based operations. As the result of these problems, many highly interactive and well designed courses in terms of their content, need an immediate feedback system (i.e., on demand) than can help students solve their computer and operating problems. After teaching beginning chemistry to over 2000 students using web-intensive materials, the most often asked questions from the students are concerned typically with problems that have arisen as the result of some type of system or web-delivery upgrade that inadvertently modifies a question in one of the learning modules. Two common types of problems that result from system upgrades are changes in set tolerance limits and in data entry formats. Other types of problems/questions are related to differences in operating systems, web browsers, and related support and other software that is installed on the students' computers, which may interfere with the interactive web-tools. Three examples of problems that can arise are an out-of-date brower's inability to recognize/display special characters correctly, security software/protocols not allowing certain types of operations (e.g., the use of macros in Excel-base emulations), and university imposed restrictions on the use of certain types of interactive programs. Even the more computer literate students often need help with some of these types of problems. For large beginning chemistry classes (i.e., containing 150-200 students or more) system and operating problems occur on a daily basis and if they cannot be addressed promptly, students become frustrated very quickly. For most first time web teachers, it is the system and operating problems that are most problematic, especially if they do not possess a high level of hardware and software expertise or do not have available university support staff that can help them. Even with appropriate, university support many student problems occur at nontraditional work times. This later problem is discussed in greater detail later in the Chapter.

As noted previously, the example course being considered, first term chemistry, is composed of several mathematical problem sets. Each of these contains a fairly large size database to increase their variability. Additionally, each problem contains an on-demand interactive "Help" link. The importance of this arrangement, in terms of providing student immediate feedback to their problems solving efforts, has been discussed above. In the past, the unfortunate downside to this highly interactive format is that a significant amount of effort is needed in setting it up, as well as in monitoring and making necessary corrections, each time the university updates/upgrades its licensed web-delivery platform. After a major update and prior to the start of a course, the overall time demands have averaged several hours per day for several weeks to check all of the individual problems and corresponding database for possible problems. This has occurred once per year and sometimes more frequently over the last six to eight years as web-delivery platform changes are made. It is these time demands that are seldom discussed and are often the most problematic issues for many who are responsible for designing and/or maintaining web-interactive courses. Although the exact solution to problem related to web-delivery and other platform updates can vary significantly for different colleges and universities, the authors' internal solution has been to abandon the university web-delivery platform completely and to setup their own servers, which operate under an open source Linux operating system in combination with an open source web-delivery platform (i.e., Moodle). The pros of this latter arrangement are: 1) greater design freedom, 2) less maintenance arising from unwanted general updates, and 3) greater ability to use various types of highly interactive custom developed software. However, the cons are: 1) a much higher level of computer/software skills are needed in setting up, operating, and maintaining the system, 2) unrestricted web-access is required, which is sometimes not possible at all user locations, and 3) all software and hardware problems must be handled directly and in a "near-immediate" fashion. A web-server being down for only a few minutes can result in a number of e-mails from angry students. As such, it is important to have built-in hardware and system redundancy that is automated or can be handled remotely. It also is important to recognize that some of the built-in backup features of current versions of Moodle are not fully functional and that other means of maintaining the course databases and student files are needed to recover from a catastrophic system failure. None of these problems is insurmountable, but they require a greater than average level of hardware and software knowledge.

Day-to-Day Questions

E-mails are to an online class what office hours are to a traditional class. They represent the human touch and a quick response time 'personalizes' the class for the online student. Unlike content related questions, when operating and system problems arise; students expect a mechanism to obtain help almost immediately, which as noted previously, can create a major timing problem. Many of the students' online activities and hence questions, occur at times that do not mesh well with most faculty schedules or university computer support staff. Figure 4 contains a representative composite graph of the percentage of e-mails received vs. the time-of-day during the Winter 2005 and Fall 2006 Quarters. This graph was constructed using representative subsets of 300 randomly selected e-mails for each of the two quarters out of the more than 3000 received. As this graph shows, there is a relatively small period of time when there are no e-mails sent/received for both years, this time period is between 3:00 and 7:00 a.m. Similar observations have been made in previous years and other terms.

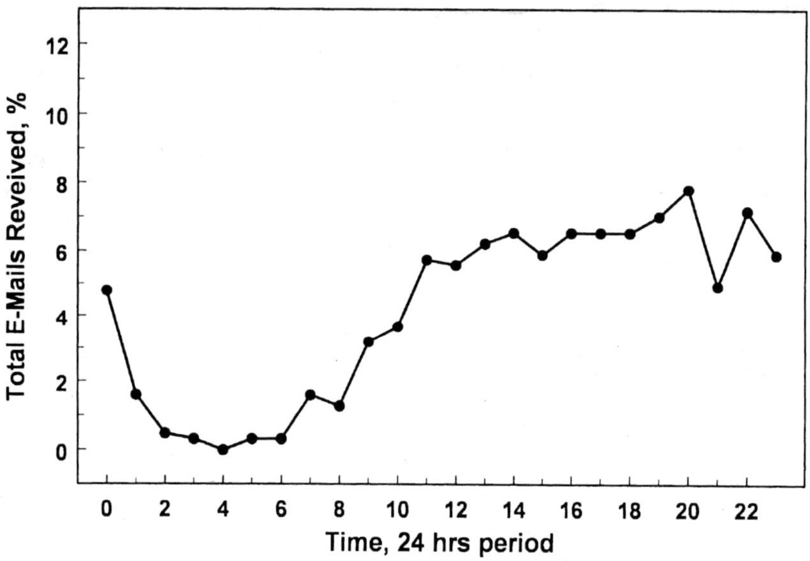

Figure 4. E-mails from students Winter term 2005 and Fall terms 2006.

Also as mentioned previously, the study and e-mail habits of students do not mesh well with the more traditional hours of a 9 – 5 job nor the schedules of most faculty members and university staff. Additionally, students expect a timely response (i.e., within a few hours) to their messages irrespective of the time-of-day. For the first half of the day (i.e., between 12:00 a.m. to 11:59 a.m.), on the average, for the two quarters, 11 to 12 e-mails were received from students, but e-mails were received at three times this rate during the afternoon and evening hours (12:00 noon to 11:59 p.m.). These same trends were observed when reviewing the e-mail messages for other weeks throughout the quarter as well as during other quarters.

Since most students expect their e-mails to be responded to in a few hours, it is difficult for a single individual to handle a large class. For the particular online class being discussed, it is fortunate that two individuals are involved in running the class. The professor in charge, who is responsible for the lecture and other content materials, and an assistant, who is responsible for maintaining the online day-to-day operations of the course, which includes the first response to e-mails, especially if they are not related to course content. Even with this arrangement, time commitments can be extremely large unless a reasonable operating plan is developed and shared with the participating students. After trying many different approaches, to date, the most practical of these is to set aside regularly scheduled blocks of time to review and to respond to student e-mails. Based on an average response time of about 4 minutes per e-mail, the total time commitment to respond to 300 e-mails per week has averaged 20 hours.

Students are given a 'schedule of e-mail response times, which is posted on the course homepage. The schedule lists times when students can expect a reply to their e-mails within 45 minutes of the time they are received by the course assistant. Students are told the first day of class that the times will be posted and that, because of meetings or other potential time conflicts, these are subject to change. Any changes known in advance will be posted and one of the students' course responsibilities is to check the schedule weekly. Typical time blocks currently being used are 9:30 – 11:30 a.m.; 1:00 – 3:00 p.m., and 7:30 to 9:30 p.m. Even though it is difficult to change student study habits and when they send e-mails, a well defined schedule, which indicates when they can expect to receive a quick response, cuts down on 'panic' e-mails (i.e., repetitive e-mails where a student may send several message that relate to the same question/problem).

Summary

Over the last decade there has been a rapid and increasing growth of the Internet and its application as an important teaching tool. Many educators are developing and using Web-based courses successfully in both the completely distance learning and the content-enhanced on-campus modes at small, intermediate, and large universities throughout the Nation. If properly designed and maintained, these courses enhance the learning experience for students. However, if they are done poorly both students and educators suffer. It is important to emphasize that the unanticipated operational problems are often more difficult to manage than the course content problems. Many first time online educators underestimate the amount of time needed to maintain online courses and to respond to the needs of the students in a timely fashion. The ongoing time commitment needs to be understood from the very beginning when a course is first in the conception stage. Otherwise, irrespective of the quality of the content and the knowledge of the instructor, the successful operation of the course will be overshadowed by student discontent in terms of non content related problems.

References

1. Gilpin, R. K 8th Sloan-C International Conference on Asynchronous Learning Networks, Orlando, FL, Nov. 2002: The Sloan Consortium. http://www.sloan-c.org/conference/proceedings/2002/track6.asp.
2. Shank, P. *Dist. Educ. Rep.* **2005**, 5-7.
3. Solch, J. G., Gilpin, R. K., and Wagel. D.J.Pittcon 2003, Orlando, FL, March 2003.: Pittsburg Conference on Analytical Chemistry and Applied Spectroscopy. http://www.pittcon.org/archive/2003.
4. Kortemeyer, G. *Am. J. Phys.* **2006**, *74*, 526-536.
5. Cheng, K.K., Thacker, B.A., Cardenas, R.L., Crouch, C. *Am. J. Phys.* **2004**, *72*, 1447-1453.
6. Gilpin, R. K., Gilpin, C. S., and Solch, J. G. 11th Sloan-C International Conference on Asynchronous Learning Networks, Orlando, FL, August 2005. The Sloan Consortium. http://www.sloan-c.org/conference/proceedings/2005/index.asp.
7. Gilpin, C. S., Solch, J. G. and Gilpin, R. K. Pittcon, 2006, Orlando, FL, March 2006. Pittsburgh Conference on Analytical Chemistry and Applied Spectroscopy. http://www.pittcon.org/archive/2006.
8. Iskander, M. F. *IEEE Trans. Microw. Theory Techn.* **2002**, *50*, 1015-1020.
9. Month, M. *Am. J. Phys.* **1999**, *68*, 556-570.
10. Conceição, S.C.O *Adult Educ. Q.* **2006**, *57*, 26-45.
11. Yang, Z., Liu, Q. *Comput. Educ.* **2005**, *48*, 171-184.
12. Humar, I. *IEEE Trans. Educ.* **2005,** *48*, 664-675.
13. Harden, R. M. *Med. Teach.* **2002**, *24*, 261-267.
14. Harden, R. M. *J. Contin. Educ. Health Prof.* **2005**, *25*, 43-51.

15. Kidd, T. *Int. J. Instruct. Tech.Dist. Learn.;* http://www.itdl.org **2005**, October. (accessed Nov 17, 2006).
16. Brinkerhoff, J., Koroghlanian, C. M. *J. Educ. Comput. Res.* **2005**, *32*, 27-56.
17. Humar, I., Bulic, E., Sinigoj, A. R. *Electrotech. Rev.* **2004**, *41*, 1-8.
18. Abdelraheem, A. Y *J Interact. Online Learn.* **2003**, *2*, 1-9.
19. Savenye, W.C, Olina, Z and Niemczyk, M. *Comput. Compos.* **2001**, *18*, 371-385.
20. Gaide, S. *Dist. Educ. Rep.* **2005**, *9,* 4-5.
21. Campbell, T. *Bull. Sci., Tech, & Soc.* **2006**, *26*, 378-387.
22. Liu, Y..; http: www.itdl.org , **2006**, April. (accessed Nov 17, 2006).
23. Owston, R.D. *Educ. Res.;* **1997**, *26*, 1945-1954.
24. McGorry, S.Y. *Internet High. Educ.* **2002**, *5*, 167-175.
25. Suson, D.J., Hewett, L.D., McCoy, J. and Nelson, V. *Am. J. Phys.* **1999**, *66*, 520-523.

Chapter 16

Analytical Science Education for Chemical Laboratory Technicians

John Kenkel

Laboratory Science Technology, Southeast Community College, 8800 O Street, Lincoln, NE 68520

Educating chemical laboratory technicians in the analytical sciences is a special and unique task. The knowledge and skills of newly-trained technicians must meet the requirements of the local industry. In addition, if the training program is to obtain formal Americal Chemical Society (ACS) approval, it must also conform to ACS guidelines. In this chapter, the ACS guidelines and the approval process will be discussed. In addition, some recently-created materials published expressly for chemistry technician education will be described and the curricula of selected ACS-approved educational programs will be examined.

Community colleges in general are not known for offering courses in analytical chemistry. The chemistry courses offered in the vast majority of community colleges are for academic transfer to four-year colleges and universities and are therefore those courses that are in high demand enrollment-wise. Such courses are in general freshman chemistry and organic chemistry. However, there are a relatively small number of colleges in the United States (U.S.) that offer technical programs (two-year programs) for educating chemical laboratory technicians and, as such, find analytical chemistry at the heart of their curricula. In this chapter, we explore the standing of analytical science in these technical programs.

In order to conform to the requirements of the local industries that hire chemical laboratory technicians, special alliances or advisory committees are utilized. According to the ACS web site (*1*), an alliance is a group of representatives from industry, academia, government, and the community that share responsibility for the education and development of chemical technicians. Such groups, among other things, meets regularly to review and approve the content of the curriculum, provides internship experiences for the students, and provides professional development opportunities for faculty. The specific knowledge and skills that graduates of the program should have become clear through this high level of interaction. It is possible that the alliance will designate specific competencies that the graduates of the program should have. In that case, faculty can present a curriculum to the students that targets these competencies.

But besides the obviously important involvement of the local employing industries, technician programs often seek ACS approval of their curriculum and program. This service of the ACS is called the Chemical Technology Program Approval Service (CTPAS) (*1*) and is a function of Subcommittee D of the ACS Society Commmittee on Education (SOCED). This service, which has been in existence since approximately 1992, examines program attributes on a number of different fronts. These attributes include the stated mission and goals of the program, the operation of the alliance (especially with regard to defining the competencies), the qualifications of the faculty, the facilities and equipment, the administration support, and the student enrollment. With respect to defining the competencies, a "gap analysis" tool is available to determine the gap between what is covered in the program curriculum and what the local employing industry emphasizes.

ChemTechStandards

In 1997, the American Chemical Society published a volume (*2*) listing over a thousand competencies that chemical laboratory and chemical process technicians should master in order to be able to function in the modern high-performance workplace. Over half of these competencies were for chemical laboratory technicians. They became universally known as the Voluntary Industry Standards (VIS) for chemical technicians and are now known as "ChemTechStandards." Their development began in 1993 when ACS was awarded one of twenty-two grants from the U.S. Department of Education to develop skill standards, in this case for chemical technicians. They were the result of an exhaustive study of the day-to-day work of chemical technicians from all types of industries and all locales. It was a modified DACUM study, meaning that the study directors interviewed hundreds of technicians nationwide in an attempt to find consensus on what knowledge and skills a technician needs to have. The entire VIS database was updated in 2003 and is now on line (*3*).

While a revision is currently in the works, the standards have, from the beginning, been organized according to eleven "critical job functions," five for laboratory employability (LE) skills and six for laboratory technical (LT) skills. The six job functions relating to technical skills are listed in Table 1.

Table 1. The six critical job functions that reflect technical skills for chemical laboratory technicians.

#	*Skill*
LT1	Sampling and Handling Chemical Materials
LT2	Measuring Physical Properties of Materials
LT3	Performing Chemical Analysis
LT4	Performing Instrumental Analysis
LT5	Designing and Conducting Experiments
LT6	Synthesizing Compounds

Note the heavy orientation toward analytical science.

In turn, these critical job functions are broken down into "modules" or "key activities" categories and the competencies are found under these activities. Table 2 lists the activities found under "Performing Chemical Analysis." The associated number of competencies for each module are given in parentheses. Table 3 lists the modules found under "Performing Instrumental Analysis." Again, the number of competencies are given in parentheses. All competencies may be viewed on line (*3*).

It is clear that much of the work of chemical laboratory technicians is analytical.

Educational Materials

The United States Congress passed the Scientific and Advanced Technology Act (SATA) in 1992. This act authorized the National Science Foundation to create a funding program for community colleges for the express purpose of improving educational programs for individuals desiring to enter the nation's technical workforce. The funding program came to be known as the Advanced Technological Education (ATE) program. Fortuitously, the timing of this program's grant offerings coincided nearly exactly with the ACS Voluntary Industry Standards (VIS) project. Through ATE, there were two major projects funded that had as a major focus the creation of educational materials based on the standards. One was the "Assignment: Chemical Technology" (ACT) series of projects at Southeast Community College in Lincoln, Nebraska, and the other was the "Contextual Laboratory Curriculum for Chemistry and Technology (C_3T) project at Athens Technical College in Athens, Georgia.

Table 2. The twelve modules listed under the critical job function "Performing Chemical Analysis" and the associated number of competencies

#	*Module Title (number of competencies)*
1	Reading Analytical Methods (4)
2	Preparing Analytical Solutions (8)
3	Preparing Samples for Chemical Analysis I – Getting the Samples into the Required Form (7)
4	Preparing Samples for Chemical Analysis II – Isolating the Material to be Measured. (7)
5	Measuring pH (9)
6	Performing Volumetric Analysis I – Acid-Base Titrations (11)
7	Performing Volumetric Analysis II – Oxidation-Reduction Titrations (6)
8	Performing Volumetric Analysis III – Complexometric Titrations (4)
9	Performing Colorimetric Analysis (11)
10	Performing Gravimetric Analysis (7)
11	Performing Electroanalytical Techniques (8)
12	Workplace Experience – Chemical Analysis (5)

The Assignment: Chemical Technology (ACT) Materials

The ACT projects consisted of a series of three projects designed specifically to create educational materials for two-year chemical laboratory technology programs. The materials created included a textbook for general chemistry (*4*), a lab manual for general chemistry (*5*), an applications-oriented cd-rom for general chemistry (packaged with the general chemistry textbook), a new edition of a popular analytical chemistry textbook (including lab experiments) (*6*) , a cd-rom for analytical chemistry packaged with the textbook, and a monograph on quality in an analytical laboratory (*7*). The analytical chemistry textbook and the quality monograph were designed specifically to address analytical chemistry requirements.

While the Voluntary Industry Standards were studied closely during the development of the books, none of the books were limited to just this one source. The authors drew from job shadowing experiences at DuPont laboratories in Texas and West Virginia, a conference of practicing technicians (*8*), local alliance participants, faculty at other colleges, and the previous editions. The book takes a traditional approach to the subject in terms of coverage of classical and instrumental methods, but the emphasis is on more practical matters than textbooks designed for baccaleaurate programs.

Table 3. The sixteen modules listed under the critical job function "Performing Instrumental Analysis" and the associated number of competencies.

#	*Module Title*
1	Overview of Instrumental Analysis (2)
2	Troubleshooting and Maintainance (11)
3	Calibration (10)
4	Sample Preparation (7)
5	Principles of Spectroscopy (7)
6	Molecular Spectroscopy I – Ultraviolet - Visible Spectroscopy (9)
7	Molecular Spectroscopy II – Infrared Spectroscopy (18)
8	Molecular Spectroscopy III – Mass Spectroscopy (10)
9	Molecular Spectroscopy IV – Nuclear Magnetic Spectroscopy (10)
10	Atomic Spectroscopy I – X-ray Fluorescence (9)
11	Atomic Spectroscopy II – Emission Spectroscopy (7)
12	Atomic Spectroscopy III – Atomic Absorption (5)
13	Chromatography I – Gas Chromatography (14)
14	Chromatography II – High Performance Liquid Chromatography (13)
15	Chromatography III – Thin-Layer Chromatography (7)
16	X-ray Diffraction and Microscopy (12)

The C_3T Materials

The C_3T project has been described in the literature (*9*). A cd-rom with a complete set of forty analytical laboratory experiments was the product of this project (*10*). Included with each experiment are the following: background information, reagent preparation instructions, experiment procedures, safety information, a discussion of data treatment, tips for interpreting the results, and a correlation to the VIS. The experiments tie in closely with the VIS and the relevant VIS are made clear to the students with each experiment. Each experiment is part of a major analytical project that the students undertake. The projects are 1) Aqueous Systems: Analysis, 2) Metals: Extraction and Characterization, 3) Polymers: Synthesis and Characterizaton, 4) Pharmaceuticals: Synthesis, Delivery and Analysis, and 5) Petroleum, Characterization and Performance. Within each major project, the analytical science competencies of the VIS are front and center. For example, in the Pharmaceuticals project, four of the ten competencies found under the calibration module and seven of the thirteen found under the High Performance Liquid Chromatography module are specifically listed as being addressed as the students progress through the lab work.

Examples of Analytical Science Curricula and Courses

There are currently twelve CTPAS-approved two-year chemical laboratory technology programs in the United States. These are listed in Table 4. Eight of these responded to a request for information for this paper.

Southeast Community College, L incoln, Nebraska

SCC is on the quarter system. The program is called Laboratory Science Technology. The curriculum includes four courses that relate directly to analytical science: Analytical Chemistry For Technicians I, II, III, and Quality in the Analytical Laboratory. Analytical Chemistry I is classical quantitative analysis, including gravimetric and titrimetric methods with laboratory activities limited to three weeks of gravimetric methods and six weeks of titrimetric methods. Material included here, but missing in the typical quantitative analysis course, are physical separation methods of gravimetric analysis (loss on drying, etc.) and a rather lengthy unit on volumetric glassware. Analytical Chemistry II is spectrometric methods of analysis, including rather classical coverage (albeit emphasizing laboratory benchwork) of uv/vis, IR and atomic spectroscopy.

Table 4. The twelve colleges having chemical laboratory technician programs currently approved by the American Chemical Society.

College Name	*Location*
Community College of Rhode Island	Warwick, RI
County College of Morris	Randolph, NJ
Delaware Technical and Community College	Newark, DE
Delta College	University Center, MI
Ferris State University	Big Rapids, MI
Lansing Community College	Lansing, MI
Miami University Middletown	Middletown, OH
New York City College of Technology	Brooklyn, NY
St. Louis Community College at Florissant Valley	St. Louis, MO
Southeast Community College	Lincoln, NE
Texas State Technical College, Waco	Waco, TX
University of Cincinnati College of Applied Science	Cincinnati, OH

Analytical Chemistry III covers chromatographic and electroanalytical methods of analysis, including GC, HPLC and ion-selective electrodes. Laboratory activities for the latter two courses consist of classical experiments along with analysis of real-world samples that students themselves collect and bring to the laboratory. Unique features of courses II and III are reports on methods that the students are assigned to write after reviewing methods of analysis in books and journals. Quality in the Analytical Laboratory especially emphasizes Good Laboratory Practices (GLP) laws and the language of quality assurance.

County College of Morris, Randolph, New Jersey

CCM is on the semester system. The program is called Chemical Technology. Two analytical science courses are required, Quantitative Chemical Analysis (first semester) and Instrumental Methods of Analysis (second semester). The laboratories for both classes consist of six hours in lab per week. In the first nine weeks of the first semester course classical gravimetric analysis and titrimetric analysis are covered and this is followed by short units on spectrometric and chromatographic methods. The second semester course is split into two parts, one for spectrometric methods and one for chromatographic, electrophoretic, and ion-selective electrode methods and also includes oral projects on Good Laboratory Practices (GLP) and Good Manufacturing Practices (GMP). According to the catalog course description for this course, emphasis is placed on the comparison of methods, the collection and interpretation of laboratory data, technical report writing, and recordkeeping.

Lansing Community College, Lansing, Michigan

LCC is on the semester system. The program is called Chemical Technology. One analytical science course, titled Quantitative Analysis, is required. This course includes gravimetric and titrimetric analyses, electrochemical analysis, and spectroscopic methods of analysis; however, chromatography is not covered. The course includes two unique assignments, a 2500-3000 word paper and a 15-20 minute group oral presentation. The paper must discuss a laboratory activity that was not performed during the course. It must include description and background information as well as limitations, sources of error, and advantages and disadvantages. As preparation for writing the paper, the student must interview a person who uses the instrument/technique on the job. The oral presentation using PowerPoint slides is on one of the first six experiments that students perform in the laboratory, with no duplications. The two students in each group make their presentations to the class, including the instructor.

Delaware Technical and Community College, Newark, Delaware

Delaware Tech is on the semester system. The analytical science courses serve two technology programs, Chemical Technology and Biotechnology. However, the courses must also serve the academic transfer population; so it must fulfill the requirements for analytical courses at state universities. The two analytical science courses are Analytical Chemistry I and II. Analytical Chemistry I is classical quantitative analysis. There is, however, much emphasis on the basic tools of analytical chemistry, the evaluation of analytical data, Good Laboratory Practices, and equilibrium chemistry. This first analytical chemistry course also stresses obtaining and preparing samples for analysis, as well as gravimetric and titrimetric analysis. This is also true of the laboratory portion of the course. Analytical Chemistry II is instrumental analysis covering spectroscopy, electrochemistry and chromatography, but also electrophoretic and kinetic methods. The material in these two courses supports mainly the Biotechnology program, but the other students also benefit from realizing how broadly chemical analysis can be applied. An example of a unique laboratory activity is titled "HPLC Plumbing." As part of the procedure, students disassemble an HPLC pump to examine plungers, check valves, and seals.

Miami University Middletown, Middletown, Ohio

Miami University Middletown is on the semester system, and the program is called Chemical Technology. Practical analytical techniques are first introduced in College Chemistry Laboratory with sampling procedures, grinding, titrimetry, and spectroscopy. This is followed by a traditional Organic Chemistry sequence and Analytical Chemistry. In Analytical Chemistry the students learn classical wet analyses and instrumental techniques as well as topics not found in a traditional course. These non-traditional topics include taking apart a Spectronic-20, grinding samples, determining loss on drying, measuring insoluble matter in reagents, and sieving. This class concludes with special projects. The special Chemical Technology course and laboratory are last in the curriculum challenges students with advanced troubleshooting problems and real-world client-employee scenarios encountered by chemical technicians in the workplace. Classical instrumental methods (FTIR, UV/Vis, GC, HPLC, GCMS, and fluorometry) are used, but the course emphasis is on such industry-based topics as standard methods, identifying sources of contamination, chain-of-custody issues, quality control charts, etc. Once again students become involved in special projects near the end.

Delta College, University Center, Michigan

Delta College's program is called Chemical Technology and the college is on the semester system. Three courses with analytical science topics are required, Chemical Analysis/Instrumentation, Troubleshooting for Analytical Instrumentation, and Research Project in Science. The Chemical Analysis/Instrumentation course covers the full gamut of analytical science from sampling and sample preparation, to analytical calculations, to pH and chemical equilibrium, to wet chemical procedures, to chromatography and spectroscopy. Also included are the non-traditional topics Good Laboratory Practices and troubleshooting. The Troubleshooting for Analytical Instrumentation course continues the troubleshooting discussion and goes into much more detail. Students must be able to determine if an instrument is not functioning properly, document instrument malfunctions, interact with service personnel, perform routine maintenance operations, and perform simple repairs. In the Research Project in Science course, students decide on an appropriate topic and perform a literature search on it. They also select a problem to be investigated and collect and analyze data. A research paper is then written. Students must prepare an abstract, present technical information, list materials and safety considerations, organize and discuss the data, draw conclusions and make recommendations.

Community College of Rhode Island, Warwick, Rhode Island

CCRI is in the semester system and the program is called Chemical Technology. All chemistry topics, including analytical science topics, are covered via a four-course sequence titled Chemical Technology I, II, III, and IV. The lecture portion meets on Monday evenings (four hours) and the laboratory experiments are performed on Saturdays (seven hours). Traditional analytical chemistry is found in all four courses. In fact, the students perform simple GC and HPLC experiments early-on in Chemical Technlogy I and later spectroscopy experiments, including atomic spectroscopy experiments, are performed. The lecture course, however, consists of basic chemistry. Chemical Technology II is structured similarly. The laboratory activities emphasize gravimetric and titrimetric analysis while the lecture continues in the study of basic chemistry. Chemical Technology III completes the study of inorganic chemistry and begins the study of organic chemistry. Once again, the laboratory activities are varied and utilize numerous analytical methods and techniques, including titrimetric analysis and atomic spectroscopy, but also infrared spectrometry. The final course, Chemical Technology IV, complete the study of organic chemistry. The laboratory introduces NMR and mass spectrometry, but also provides more advanced work with GC, HPLC, etc. A special feature of the laboratory for Chemical Technology IV is that students are asked to identify "general" unknowns. This means that all analytical schemes are utilized, from simple melting and boiling points to IR, NMR and mass spectrometry.

Texas State Technical Institute, Waco, Texas

TSTC's program is called Chemical/Environmental Technology and is on the semester system. The curriculum is 67 credits over 5 semesters. The course curriculum is guided by industry partners and biennial regional validation of the ACS Voluntary Industry Standards (VIS). The competencies for the courses are derived from the VIS. Analytical sciences are addressed specifically in applied analytical chemistry, applied instrumentation I & II, environmental chemistry and advanced environmental chemistry. The applied analytical chemistry course addresses traditional, or gravimetric and volumetric, wet chemical methods, including fundamental statistics and Good Laboratory Practices (GLP). Analytical methods are taken directly from "Standard Methods for the Examination of Water and Wastewater" and American Society for Testing and Materials (ASTM) procedures. The applied instrumentation courses cover chromatography (TLC, GC and HPLC), UV-VIS, AA, FTIR, GCMS, NMR and X-Ray fluorescence. The emphasis is on sample preparation and analysis, basic instrument theory, design, and operation, and basic troubleshooting. The first environmental chemistry course addresses fundamental concepts and the special considerations for analysis of air, water and soil, including the required documentation. The laboratory component of this course utilizes LaMott field test kits. The advanced environmental analysis uses standard bench methods. Students in these courses have designed and conducted semester-long projects, including wetlands water monitoring and soil analysis projects.

Assessment Strategies

Chemical laboratory technicians must have a high level of skill with respect to bench work in an industrial analytical laboratory Therefore, it is no surprise that assessment strategies of the educational programs focus on analytical laboratory skill and technique. The very fact that some courses require six hours or more of analytical chemistry laboratory work per week (as mentioned previously for County College of Morris and Community College of Rhode Island, for example) attest to this fact. "Accuracy" is a primary factor for assessment and evaluation. Commercially available analytical unknowns, for which the instructor knows the correct percentage of analyte, are popular, since a percent error can be the basis for a grade. Likewise, instructor-prepared unknowns (for which the instructor knows the correct answer) can be used. However, the analysis of real-world unknowns is desirable, since the students get experience with samples similar to those that they will be analyzing on the job. In this case, assessment can take the form of the instructor's observations of a student's lab technique. In either case, students must also demonstrate their record-keeping skills. The procedure for real-world notebooks is outlined in the Good Laboratory Practices laws, which are covered by most programs. Hence the notebook format mirrors that required by the GLP and assessment is based on adherence to this format in many cases. In other cases, a formal laboratory report, which includes more detail but with a similar format, are submitted and graded. It should be mentioned that all programs surveyed appear to place some value on traditional assessment tools (exams, quizzes, etc), especially for the lecture portion of the courses.

Some especially unique assessment strategies deserve specific mention here.

Capstone Course

At Southeast Community College, students take a 1-credit course during their final quarter of enrollment called "Laboratory Skills Competency." In this capstone course, the students meet one-on-one with each program instructor for the purpose of evaluating a student's laboratory skills. For the analytical chemistry portion of this activity, students are asked to design an experiment to accomplish a desired end result, such as the determination of the effect of pH on a Beer's Law experiment or the effect of pH on the extraction of an analyte metal from soil. The student is asked to write a proposal for a procedure to accomplish the goal, show it to the instructor, then re-write it as many times as is required to gain the instructor's approval. Meanwhile the instructor assigns points to each proposal draft. In the end, the student performs the experiment while the instructor watches. With pen and clipboard in hand, the instructor keeps notes and records any errors (including calculation errors, glassware use errors, etc) along with points lost for each error. The student's notebook writeup is also part of the scheme.

Department Meetings

At Miami University Middletown, "department meetings" are held each week of the Chemical Technology II course. Students, as members of an industrial "department," share what they have learned, what they have accomplished, and what problems they have encountered in the past week. Group discussion provides a way to share the learning and solve the problems. This simulates department meetings in industry in which group cohesion is maintained and a ready conduit for advice from experienced colleagues (in this case, students who have already used the instrumentation or performed the experiment) is provided. The department manager (the instructor) makes announcements, reviews company and department status, and gives work assignments for the coming week. Participation in these department meetings counts as 20% of the grade in the course.

Electronic Assessment Tools: The OWL System

At the Community College of Rhode Island, the textbook used (*11*) is tied to an electronic assessment tool known as the "OWL" system, a web-based learning system for Brooks/Cole textbooks (*12*). The OWL system is both a tutorial and an assessment tool. Students log onto the system, choose an assignment, read any given information and then answer the questions. Slightly different versions of the same questions are repeatedly presented to the student until the student achieves the correct answer or simply quits. The system only records the student's best performance on each assignment. OWL thus makes it possible for any student to achieve a perfect score for one of the four exams in the course. The successful student must exercise a diligent work ethic, however, because the assignments are only open for grading a prescribed amount of time. The OWL score counts as one 100-point exam.

Assessment Strategies at TSTC

At the Texas State Technical Institute, Waco, Texas, assessment in the analytical laboratory course is based on precision and accuracy using commercially analyzed unknowns. Additionally, this course also includes a pre-lab and post-lab practicum using a statistical analysis of pipetting technique and a written lab final exam covering laboratory techniques. The questions for the lab final are given to the student at the start of the semester. In the other courses, assessment is based on the quality of the data, demonstration of proper use of the instrument, the maintenance of a laboratory notebook following standard guidelines and a formal presentation of an experiment. A capstone course is required during the last semester. Typically students select a "problem" in which they do experimental design, logistics of collecting and storing samples, waste disposal and reporting, including a formal written report, a poster, and an oral presentation.

Conclusion

A relatively small number of two-year colleges offer programs to prepare students for careers as industrial analytical laboratory technicians. These programs accomplish a task that is well-defined and have a vast amount of resources available. They are designed to focus on analytical bench skills, and utilize assessment methodologies that measure these skills. The American Chemical Society conducted projects to develop resources and the National Science Foundation awarded grants toward this effort. The colleges that have been approved by the American Chemical Society's approval service are shining examples of how to go about educating technicians. Traditional as well as unique assessment techniques emphasize laboratory skills.

References

1. ACS Chemical Technology Program Approval Service. http://chemistry.org/education/ctpas.html. (accessed December 8, 2006), click on "alliance overview."
2. *Foundations for Excellence in the Chemical Process Industries, Voluntary Industry Standards for Chemical Process Industry Technical Workers*; Hofstader, R.; Chapman, K.; American Chemical Society Education Division: Washington DC, 1997.
3. Introduction and Overview of VIS Database. www.ChemTechLinks.org/SkillStandards. (accessed December 8, 2006).
4. *Chemistry: An Industry-Based Introduction with CD-ROM*; Kenkel, J.; et. al.; CRC Press: Boca Raton, FL, 2001
5. *Chemistry: An Industry-Based Laboratory Manual*; Kenkel, J.; CRC Press: Boca Raton, FL, 2001.
6. *Analytical Chemistry For Technicians*; Kenkel, J.; CRC Press: Boca Raton, FL, 2003.
7. *A Primer on Quality in the Analytical Laboratory*; Kenkel, J.; CRC Press: Boca Raton, FL, 2000.
8. Kenkel, J., et. al. *The DuPont Conference: Implications for the Chemical Technology Curriculum*; Journal of Chemical Education: **1998**, *75*, 5, 531-534.
9. Braun, J.; White, C. *Teaching Chemistry Laboratory Skills in Industrial Contexts*; Journal of Chemical Education: **2006**, *83*, *3*, 353-354.
10. *Contextual Laboratory Curriculum for Chemistry and Technology*; Athens Technical College: Athens, GA, 2004.
11. *Introductory Chemistry: An Active Learning Approach*; Cracolice, M.; Peters, E.; Brooks/Cole: Belmont, CA. 2003.
12. OWL, Online-Based Learning. http://owl1.thomsonlearning.com/. (accessed December 8, 2006).

Chapter 17

Teaching and Learning Quantitative Analysis

Robert J. Eierman

Department of Chemistry, University of Wisconsin at Eau Claire, Eau Claire, WI 54701

This article investigates whether students are learning more deeply and whether Quantitative Analysis (QA) is being taught more effectively than a generation ago. A review of the literature shows interesting developments in the content and pedagogy used in teaching QA. However, most developments are lacking either a coherent course design or a scholarly assessment, which leaves it unclear if these developments are effective. Characteristics of the scholarship of teaching and learning are described. A course redesign and an assessment project are presented as examples of how curriculum projects can be structured and assessed to enable conclusions to be drawn about their effectiveness.

Are students learning QA more deeply? Is teaching in QA more effective than a generation ago?

There have been many fascinating developments in teaching QA in recent years. It is valuable to reflect upon whether students develop a deep, connected and multi-faceted understanding of the content and skills of analytical chemistry and what aspects of various curriculum designs contribute to the student learning.

The academic community has developed scholarly methods designed to answer complicated questions. Effective scholarship has clear goals that are addressed by a well-prepared scholar and are studied using appropriate methods. The results must be communicated effectively, must be accompanied by reflective critique and the significance of the results must be judged by peer experts *(1)*. In the 1700's and 1800's, chemists developed methods for designing, carrying out and communicating studies that produced evidence to support conclusions about chemical systems. They developed systems with high standards that have enabled chemists to answer complicated questions about the chemical world. Chemical education researchers are now working to find effective methods that will answer such questions about issues of teaching and learning *(2)*. These systems must have the same rigor and high standards as those in chemistry research.

This chapter includes a review of QA curricular design literature that shows how the teaching of QA has evolved in the past 25 years. A discussion of some systematic methods for course development and assessment is presented. Finally, a project is described that illustrates how those methods work.

A Review of the Development of QA Curriculum Over the Past 25 Years

A literature search on QA course curriculum reveals how it has developed and evolved. The search finds three primary types of articles: analytical chemistry philosophy, applications of pedagogical methods and theme-based lab curricula. In addition, two surveys about teaching analytical chemistry provide insight into QA course evolution. A discussion of the role of textbooks in QA course development is also presented.

Analytical Chemistry Philosophy

Every few years leaders in analytical chemistry, either individually or in groups, publish articles describing how the field of analytical chemistry "works" and suggesting how to incorporate that into analytical chemistry coursework. These articles have had varying levels of impact on curricular developments in the subsequent years.

In 1979, an ACS symposium on the Status of Teaching Analytical Chemistry was summarized in a series of short articles *(3)*. Questions were raised about teaching many techniques vs. teaching careful, accurate lab skills and/or how techniques work. Concern was also expressed about the balance of wet chemical vs. instrumental techniques. It was suggested that textbooks should include more information on the process of analytical chemistry, on the selection of the best methods and that coverage of equilibrium principles and titrations should be reduced. Laitenen published an article describing the process of analytical chemistry as requiring ingenuity and intuition in addition to an orderly and systematic approach *(4)*. Pardue proposed teaching analytical chemistry using a unified approach that takes into account the problem-based (vs. sample-based) nature of the discipline *(5)*. A joint industry-academic symposium concluded that analytical chemistry students need a broad background in analytical sciences, need to understand instrumentation and need to maintain intense professional activity in order to succeed *(6)*.

In 1995, Christian described the history of QA textbooks and topics and described role-playing groups as an effective teaching method *(7)*. An NSF report on curricular developments in analytical chemistry *(8)* stated that students need to know more about the fundamentals of analytical measurement and the scientific method. Problem-based learning was touted as a method to bring about such understanding. In a less-cited article, Valcarcel called for design of the QA curriculum from the bottom up, starting with measurement fundamentals, moving to how qualitative and quantitative measurements are made and culminating with analytical problem solving *(9)*.

Applications of Specific Pedagogical Methods in QA

The QA pedagogies that have been published include use of lab projects, cooperative groups, role-playing and problem-based learning. Each is described, though overlap of these topics occurs frequently.

Lab projects in QA probably go back to the origins of the QA course and have been used exhaustively (*10, 11, 12, 13, 14*). Typically they are included in courses to enable students to apply analytical methods to "real" samples, to connect course material to students' interests and to have students engage in the process of analytical chemistry. Students are given various levels of resources, responsibilities and requirements depending on the project goals. Projects are frequently used as culminating events in the course with students presenting their results in talks or posters.

Cooperative learning philosophy and methods were developed in K-12 teaching and subsequently moved to the college level *(15)*. In cooperative groups, students work together actively to learn content or carry out tasks. For successful function, the assessment of learning in cooperative groups must include both individual and group accountability. Cooperative groups have been used in QA lecture to encourage students to be active and to work together and learn communication skills *(16, 17)*. Improvements in problem-solving skills and students' attitudes were reported.

A special case of cooperative learning in lab was reported by Walters *(18, 19, 20, 21)*. Students work in cooperative role-playing groups during lab in a couple of analytical chemistry courses. Each student has a role, designed after industrial chemists, and each role defines the responsibilities of that student. The experiments are designed so that positive interdependence and good management, which typically lead to good results, are rewarded with good grades. Each four-student group has a set of equipment to use in solving the problem posed by each experiment and the students rotate through the roles in subsequent experiments. The instructor serves as "upper management", setting the tasks and receiving the reports.

Problem-based learning (PBL) is another group-centered activity in which students work together to solve problems. PBL is done in the lecture or lab, with student groups working to solve problems. In one QA course students have a PBL lecture section, where they work problems in small groups until all groups have gained understanding, they then have a summary discussion and move to the next subject *(22)*. In lab, students work in different small groups on long-term projects, such as analyzing all components of coffee using liquid chromatography. In two other courses *(23, 24)*, students define lab problems, research and select methods and carry out analyses of assigned samples. PBL helps students develop responsibility, organization, teamwork and communication skills and an understanding of the scientific method.

NOTE: Activities for an approach called Process Oriented Guided Inquiry Labs (POGIL), are just beginning to be developed for analytical chemistry courses, and have not yet appeared in publications.

Theme-based QA Laboratories

Several articles have been published that describe analytical chemistry labs that are designed around themes. Themes are a creative way to focus learning in the lab. The articles did not describe the lecture part of the course.

A 250-gallon, salt-water aquarium, including live sea creatures, served as the source of all lab samples for a QA course *(25)*. Students measured and monitored levels of nitrate, ammonia, dissolved oxygen, salinity, pH, sulfate and metal ions. Analytical methods used included acid-base, complexometric and redox titrimetry, gravimetric analysis, spectrometry and potentiometry. Students took turns monitoring the analytes, observed trends in their levels and learned to correlate levels with the health of the aquarium. Student response to the laboratory was very positive, although no formal assessment was done. Another course was modified in a related way *(26)*. The first half of the course was designed in a traditional format with a mix of wet and instrumental methods. In the second half, students analyzed water samples from a fresh water fish aquarium as well as lipid and moisture levels in the fish. Chemometric methods were introduced as well. Students said they learned new skills due to analyzing the "real" samples as well as group, leadership and problem-solving skills.

Lead, in a variety of forms and matrices, serves as the only analyte for an instrumental analysis course *(27)*. During the first 2/3 of the course students work in groups to analyze lead using spectrometry, fluorescence quenching, GFAAS, NMR, cyclic voltammetry and ASV. Students then select and use a method to extract and analyze Pb in soil samples collected from the area near the university. Students work with an external client to define an environmental lead problem that can be solved using the results. The students are actively engaged in the lab and in interactions with the clients, which results in a rich learning environment.

A QA course was restructured using advice from an advisory group with members from academia, national labs and industry to include fundamentals of analysis, including good laboratory practices, as well as studies in contemporary analytical science *(28)*. Students undergo certification, perform quality assurance experiments and use a wide range of chemical and physical analysis systems. Student perceptions were favorable and they perceived that their experiences would be helpful in their careers.

The final theme-based lab was developed in western Australia where there are many companies employing chemists to discover and analyze minerals *(29)*. The lab was designed around a variety of mineral analyses. Students learn a great deal about sample preparation as well as wet and instrumental methods.

It is important to note that few of the publications include substantial assessment of student learning and attitudes. Some authors have carried out formal, structured assessment, but in most cases, the assessment was narrow and not focused on the overall goals of the curriculum changes.

Surveys of Teaching Analytical Chemistry

Two surveys have been published regarding issues of teaching analytical chemistry *(30, 31)*. The first survey asked QA professors about course logistics, topics taught in lecture and experiments taught in laboratory. The second survey asked about the faculty member, the QA lecture and about the use of PBL in QA. The surveys were published about 15 years apart and 109 and 62 responses were received for the surveys respectively.

A few conclusions can be drawn despite differences in survey structure:

1) In the first survey, 31% used Skoog *(32)* and 24% used Harris *(33)* as their textbook, where in the second, over 50% used Harris and 25% used Skoog.
2) A similar set of QA lecture topics is reported in each survey, but there is a clear increase in percentage of courses that include more spectrometry and separations (GC and HPLC) and less wet methods (gravimetry, complexometric and redox titrations) in the second survey. The analytical method, not included in the first survey, is included in 78% of courses in the second.
3) In lab, the first survey reported that acid-base, redox and complexometric titrations along with spectrometry experiments were done in over 80% of courses. The second survey did not ask about experiments, but did report that 69% use Thorn-Smith samples for at least one experiment and that 69% use some form of a group laboratory project.
4) In the second survey, 44% report using PBL in their classrooms, 71% reported reading Analytical Chemistry (AC), 63% report reading the

Journal of Chemical Education (JCE) and none report reading the Chemical Educator or the Journal of Research in Science Teaching.

A survey done by the author of 86 QA course web pages in two eastern states (CT and MD), a midwestern state (WI) and two western states (OR and WA) shows that QA is usually a 4-credit course with 2 or 3 hours of lecture and 4 to 6 hours of lab per week. Course structure is similar from region to region.

The Role of Textbooks in QA Curricular Design

Chemistry instructors typically use a single textbook in teaching a course. In QA, two textbooks dominate the market, which have many similarities in content and organization *(32, 33)*. For many instructors, particularly early in their careers, the textbooks serve as a de facto curriculum design. Are QA textbooks effective designs for courses? Do textbook authors design the books for that purpose?

These issues were discussed with Daniel C. Harris, author of the most widely used QA textbook *(34, 35)*. Harris describes the following features that he has tried to incorporate in the book.

- QA should be presented with enthusiasm and in an interesting manner.
- Explanations should be clear and complete with an optimal balance between completeness and length.
- Topic discussions move from concrete examples to abstract concepts.
- Problem solving is taught by example and by providing practice problems with feedback.
- The book covers a wider range of topics than is included in a typical course, which enables instructors to select their desired content.

Over two decades of the Harris book, coverage of spectrometry (including mass spectrometry) and chromatography have increased along with sampling and the analytical process. Reductions have occurred in electrochemistry, equilibrium and gravimetry and lab experiments have been moved to a web site to shorten the book. These trends reflect the QA survey results discussed earlier. Harris believes that possible improvements include integrating instrumentation better, more guidance on the analytical process and support for problem-based learning and problem solving. In addition, Harris thinks that two topics, the context of chemical analysis and biological analysis, should receive more emphasis.

Textbooks are written for instructors to use in many different ways. They must be flexible with a broad range of topics, but must balance depth of coverage with breadth. Where possible, authors avoid focusing on a specific course design, which makes a textbook less marketable. Based on user surveys of the Harris book, the first half of the book (tools, error and statistics, wet chemistry and equilibrium) is used in the chapter sequence more frequently than the second half (spectrometry, chromatography, electrochemistry). Textbook topics are chosen from trends in the analytical chemistry literature and from topics instructors want to include in their courses. The conclusion is that textbooks cover a wide range of topics with explanations of concepts and systems, and guidance and practice in problem solving, but leave course design to the instructors.

In summary, a review of the literature shows that the curriculum in QA is guided by analytical philosophy, includes wet chemical analysis, an increasing emphasis on instrumental analysis and is increasingly being taught in a way to provide students experience with an understanding of the process of chemical analysis. A range of interesting pedagogies are used in teaching QA, but formal assessment is not part of most QA curriculum literature. Textbooks are written to provide a range of topics, clear, complete explanations of analytical concepts and guidance and practice in problem solving, but not as designed courses.

Systematic Design and Assessment of QA Curriculum

The review of QA curricular literature shows that much work has been done through the years, but there is not a great deal in that work that helps to answer what approaches lead to deeper understanding of QA. To find answers, systematic curriculum design of QA and scholarly assessment of the curriculum must be done. The chemistry community must use scholarly methods to answer these rich, complicated questions, just as it uses scholarly methods to answer chemistry research questions. Systematic curricular design and scholarly assessment will be described briefly and a QA course redesign/assessment project will be described that serves as an example of how they work.

Systematic Curricular Design and Scholarly Assessment

In the past several decades, sophisticated models of how people learn have been developed through advances in educational psychology, brain physiology and discipline-based pedagogical research *(36, 37)*. These models can serve as the basis for designing effective pedagogical methods. Processes for carrying out efficient, coherent design of curriculum have been described in detail *(38, 39)*. The processes ensure careful definition of content and skills to be learned, design of appropriate assessment of learning and creation of student and instructor activities that focus on enabling deep learning. Deep learning is defined as developing connected and multi-faceted understanding of the material. Most of the QA literature describes curriculum that was developed with a specific focus, e.g. change the lab without the lecture. The change of one aspect of the course may be beneficial for that part, but may not improve the overall learning.

The scholarship of teaching has been defined *(40)* as one of four types of scholarship in academia, one of which has evolved into the scholarship of teaching and learning (SoTL). The characteristics used to assess such scholarship have been defined to include the following *(1)*:

1. Has clear goals (well-defined, feasible).
2. Carried out by a prepared scholar (command of field).
3. Appropriate methods are used to carry out the investigations.
4. Is judged by significance of results.
5. Is communicated effectively.
6. Is accompanied by reflective critique.

A scholarly assessment of the impacts of curricular redesign on student learning should show these characteristics and is the best way to answer the question of whether QA teaching and learning are improving. The QA literature review shows that curricular assessment has been done sporadically and often does not display the above characteristics.

The design of appropriate systems to assess curriculum has been described *(41, 42, 43)*. Questions are asked regarding an issue of teaching and learning. A range of types of evidence necessary to answer the questions and tools needed to gather that evidence must be selected or designed. The evidence is gathered, analyzed, conclusions drawn, reflected upon and communicated to the community in ways that enable peer review and judgment of the significance. This sort of scholarly process is the best means to build a body of knowledge that can answer complicated questions like those asked at the top of this article.

An Example QA Design and Assessment Project, The New Quant Project

The following is a QA course redesign and assessment project carried out by the author that shows many of the characteristics described above. It is not the definitive study in QA design and assessment, but rather represents the type of study that will enable the community to build a body of knowledge that will lead to more effective teaching and deeper learning of QA.

Principles used in the course redesign

The three stages in a backward design *(38)* are:

1. Identify desired learning outcomes
2. Determine acceptable evidence
3. Plan instruction and learning experiences

To identify desired results, the instructor selects contents and skills to be learned and prioritizes them as:

- Enduring understandings (big themes)

- Important to know and do
- Worth being familiar with

To determine acceptable evidence the instructor decides what a student who has mastered the course content will "look like", i.e. will know and be able to do. Assessment activities are designed that will enable the instructor to recognize student mastery. This is the "backwards" part because the assessments are designed before the learning experiences. The continuum of assessment methods is also defined, from least to most formal:

Informal Checks → Observation/dialogue → Quiz/Test → Academic Prompt → Performance Task

Notice that tests are in the middle of the continuum with prompts and projects being the higher forms of assessment in terms of probing deep understanding.

Finally, instructional activities are planned to maximize students' abilities to achieve deep understanding and succeed on the assessments.

Description of the curriculum including assessments

Defining "enduring understandings" in QA was an interesting challenge. At first, seven or eight understandings connected to analytical methods were envisioned. However, it was decided to tie the understandings to big themes of science to enable students to connect this course with other science courses and experiences. One list of big themes is seen below:

Unifying Science Concepts and Processes [3]

- Systems, order, organization
- Evidence, models, explanation
- Constancy/change, measurement
- Evolution, equilibrium
- Form, function

Upon reflection, it became clear that no more than three or four enduring understandings could be addressed. Three understandings selected that encompass the content of QA are Measurement, Error and Models. The course was designed in three units, each with a series of lecture and lab activities focused on the enduring understanding. The units have a common structure and include a range of assessment types, including academic prompts (lab projects).

In the Measurement Unit, all activities center on the concept of measurement (see Table I). Focused, short textbook reading assignments are given from various chapters in the book, since its sequence is different from the lecture sequence. Much of the practice with lecture concepts consists of homework using problems from the book. The lab experiments focus on the same measurement concepts described in lecture. Lab write-ups are a combination of short answers and calculations of data collected in lab. A one-hour exam is given at the end of week four, graded and returned early in week five. Students then complete a laboratory project using a self-designed process.

Table I. The Measurement Unit

Week 1	Week 2	Week 3	Week 4	Week 5
Lecture Topics: Basics of Measurement (Standards and Comparison)	Chem. Measurement (Analyte charac's, chem. std's, calibration)	Wet Chemical Analysis (Titrations and gravimetric)	Instrumental Analysis (Spec. and potentiometric)	Start Error Unit
Lab Experiments: Glassware calibration	Volumetric Aspirin	Spectrometric Aspirin	Potentiometric Fluoride	Measure-ment Project
Assessments: Homework	Homework Lab write-up	Homework Lab write-up	Hour exam Lab write-up	Lab write-ups

In the Measurement Project (Week 5), students work with a partner to answer an instructor-provided question using a self-designed procedure. Students are given lab resources and two lab days to gather data. The report is due one day after the second lab period. The question for the Measurement Project is:

*Empirically does one tablet of **Tums** or one dose (1 tsp) of **Phillips** neutralize more acid, or are they the same?*

Students use measurement and acid-base titration ideas to answer the question.

The Error and Models units (see Table II) have a similar structure including experiments and a culminating project that focus on the theme.

Table II. Topics in the Error and Models Units

The Error Unit	The Models Unit
Experimental error (significance, types, sources and measurement)	Levels of conceptualization (pictorial, symbolic, mathematical)
Making data-based decisions (appropriate use of statistitics to make decisions)	Solution chemistry (dissolution, acid/base, complexation, redox)
Error in chemical analysis (wet chemical, instrumental)	Models of Instrumental Analysis (spectrometry, ISE's)

Implementation and assessment of the curriculum

In this project, the curriculum was assessed focusing on two central questions (see below). Evidence of student opinion and performance and instructor reflections was gathered before and after curriculum implementation.

The central questions for evaluating the QA curriculum are:

1. *Will students develop a deep understanding of quantitative analysis by following a curriculum created using a backward design?*
2. *What factors influence the learning?*

The types of evidence gathered to answer the question include:

- ACS standardized final exam (AN01 given before and after)
- Review final project poster reports and scores before and after
- Instructor reflections
- Student surveys, 2 times/semester, SALG online (http://www.wcer.wisc.edu/salgains/instructor/)
- Focus Group (Six students met three times with a research student to answer questions; sessions were audiotaped, transcribed and analyzed.)

Four conclusions and supporting evidence

1. ACS standardized exam results are about the same.

The exam was given before and after implementing the New Quant curriculum (see Table III). The average score after is about one question below that of before and both are near at the national norm. It is concluded that the new curriculum did not have a significant impact on student's ability to score on a standardized, multiple-choice exam. This is not a surprise since the curriculum changes are focused on building lecture and lab connections and science process skills. In surveys, students indicated that they heartily dislike taking the ACS exam as a final test, presumably since its structure is different from the other exams in QA. However, student scores on the ACS exam do correlate well with their overall course score (R^2 = .782).

Table III. Scores on the ACS Standardized Quantitative Analysis Exam

	Ave.	Std Dev
Before	29.4	6.2
After	28.2	6.8
National	28.5	7.6

Scores are number correct out of 50 questions.

2. The unit structure focuses content and strengthens the lecture/ laboratory connection.

Instructor reflections The unit theme was referenced essentially every day in lecture and lab. I used more lab examples in lecture and vice versa. The lab experiments, particularly the write-ups, focused on the unit theme. The overall course content included less detailed calculations, but included better connected ideas. The textbook and course topics sequences are different. Short reading assignments were given that closely relate to the upcoming lecture. The lab/lecture connection was much stronger.

SALG Survey results Students responded favorably to "How the class activities, labs, reading and assignments fit together" (3.73/5) and to "How parts of the classwork, labs, reading or assignments related to each other" (3.69/5). They responded particularly well to questions about how much they learned about measurement and error, but not so well to models.

Focus Group comments When asked what helps their learning students consistently mentioned "lab examples in lecture" and "the structure of the units is similar". They sited "the lab/lecture connection" as a strength of the course.

3. The separation of "measurement" and "error" is an improvement.

Instructor reflections The typical topic sequence in QA has a brief introduction to chemical analysis followed by a discussion of error and statistics with subsequent discussions of a range of methods. In the New Quant curriculum, the entire first unit is devoted to developing an understanding of how the various chemical measurement methods, wet and instrumental, work. When the difficult topic of error is discussed in the second unit, students have a good foundation of how the measurements work. They are better able to handle the error issues because they know how the measurements are supposed to work.

SALG Survey Results Students consistently responded very favorably that they "think they understand error and making data-based decisions" (4.02/5) and that they will "remember and carry error into other aspects of their lives" (3.82/5). Although I have no survey results from previous semesters, anecdotally these results represent a positive shift in understanding of error.

Performance Measures Two comparative performance measures regarding error do not support this conclusion strongly. Student scores on the ACS Exam section on Data Evaluation and Error Analysis were about the same before (67%) and after (64%), but both exceeded the National average (58%). Student scores on the Final Project Report section on statistical analysis and conclusions were also similar before and after. Although student performance on these two tasks did not improve, performance is strong and students seem more comfortable and relatively proficient in measurement and error.

4. The lab projects expand the curriculum to include important science skills and are a good final assessment.

Instructor reflections In the projects, students are forced to behave independently and use good science practices much more than in the typical experiments. They have to ponder and plan, ask and answer questions regarding what to do and how, negotiate decision points and recognize when to proceed and when to stop. Their cooperative and communication skills are taxed. Because there are three projects, they get suggestions and feedback about these skills when their reports are returned.

SALG Survey Results Students responded very favorably to "How much has this class added to your skills in" making chemical measurements (3.89/5), working independently in the lab (4.00/5) and critical thinking (3.90/5). They also say they will "remember and carry into other aspects of their lives" error (3.82/5), problem solving skills (4.02/5) and lab technique (4.09/5).

Focus Group Comments Students commented that the projects helped them "learn to do science" and "gain self confidence." One said "The projects are a good application of learning, we put in a lot of thinking." Another student said "We had to solve the problem, we couldn't just follow the procedure."

Performance Measures Scores on the final project reports were used as a measure of deep understanding of QA. The final projects required students to work independently to design and carry out a chemical analysis. Students apply much QA content and many analytical skills in completing the project. The report format and scoring rubric were identical.

Project scores from two spring semester classes of the same size (n = 44) taught before and after the change were compared. The average score after was about 2% higher, and the difference is statistically significant at the 70% confidence level. The standard deviation was twice as large after than before indicating a larger spread of scores. There were 13 scores of 95% or higher after and only 2 before. I conclude that the performance on the final project after the change was slightly better than before.

The project scores after correlated better with the overall course scores (R^2 = 0.644 vs 0.216). The projects are a more integral part of the course structure so the scores are a good reflection of student achievement.

The above project is not without flaws, but it has the characteristics of a scholarly curriculum design and assessment project needed to identify methods that create deep learning in QA. The course design is built on published learning and curriculum models and the assessment is structured to gather a range of evidence using appropriate methods to answer the focus questions.

The culture of analytical chemistry must change to demand scholarly studies to demonstrate the most effective methods of teaching in QA.

References

1. Glassick, C.E.; Huber, M. T.; Maeroff, G. I. *Scholarship Assessed*; Jossey-Bass: San Francisco, CA, 1997; pp. 22-27.
2. Coppola, B. *Interdisciplinary Sciences at the Interface of Education*; Talk at UW-System Chemistry Faculties Meeting, La Crosse, WI; 2000.
3. Dolbow, K.E. *J. Chem. Ed.* **1979**, *56*, 96-99.
4. Laitenen, H.A. *Anal. Chem.* **1979**, *51*, 1889.
5. Pardue, H.L.; Woo, J. *J. Chem. Ed.* **1984**, *61*, 409-412.

6. Borman, S.A. *Anal. Chem.* **1985**, *57*, 278A-285A.
7. Christian, G.D. *Anal. Chem.* **1995**, *67*, 532A-538A.
8. Kuwana, T. *Curricular Developments in the Analytical Sciences*; NSF: Washington, DC, 1998.
9. Valcarcel, M. *Anal. Chem.* **2001**, *73*, 333A-334A.
10. Roscher, N.M. *J. Coll. Sci. Teach.* **1973**, *2*, 33-34.
11. Neman, R.L. *J. Coll. Sci. Teach.* **1983**, *13*, 16-18.
12. Atkinson, G.F. *J. Chem. Ed.* **1984**, *61*, 413-414.
13. Eierman, R.J. *J. Chem. Ed.* **1998**, *75*, 869-873.
14. Marine, S.S. *J. Chem. Ed.* **2003**, *80*, 366.
15. Johnson, D.W.; Johnson, R.T.; Smith, K.A. *Cooperative learning: increasing college faculty instructional productivity*; George Washington University: Washington, DC, 1991.
16. Ross, M.R.; Fulton, R.B. *J. Chem. Ed.* **1995**, *71*, 141-143.
17. Wright, J.C. *J. Chem. Ed.* **1996**, *73*, 827-832.
18. Walters, J. P. *Anal. Chem.* **1991**, *63*, 977A-985A.
19. Walters, J. P. *Anal. Chem.* **1991**, *63*, 1077A-1086A.
20. Walters, J. P. *Anal. Chem.* **1991**, *63*, 1179A-1191A.
21. Jackson, P.T., Walters, J.P. *J. Chem. Ed.* **2000**, *77*, 1019-1025.
22. Wenzel, T.J. *Anal. Chem.* **1995**, *67*, 470A-475A.
23. Ram, P. *J. Chem. Ed.* **1999**, *76*, 1122-1126.
24. Wilson, G.S.; Lunte, C. E. *Anal. Chem.* **1999**, *71*, 677A-681A.
25. Hughes, K. D. *Anal. Chem.* **1993**, *65,* 883A-889A.
26. Houghton, T.P.; Kalivas, J.H. *J. Chem. Ed.* **2000**, *77*, 1314-1318.
27. Fitch, A.; Wang, Y.L.; Mellican, S. *Anal. Chem.* **1996**, *68*, 727A-731A.
28. Perone, S.P.; Pesek, J.; Stone, C. *J. Chem. Ed.* **1998**, *75*, 1444-1452.
29. Phillips, D.N. *Anal. Chem.* **2002**, *74,* 427A-430A.
30. Locke, D.C.; Grossman, W.E.L. *Anal. Chem.* **1987**, *57*, 829A-835A.
31. Mabrouk, P.R. *Anal. Chem.* **2002**, *74*, 268A-274A.
32. Skoog, D.A.; West, D.M. *Analytical Chemistry: An Introduction, 4th ed*; Saunders: Philadelphia, PA, 1986.
33. Harris, D.C. *Quantitative Chemical Analysis*; Freeman: New York, 1982.
34. Harris, D.C. *Genesis and Evolution of a Textbook*; Talk at Walter Harris Workshop, Edmonton, Canada, 2005,
35. Harris, D.C. *Personal Communication.*
36. *How People Learn*; Bransford, J.D.; Brown, A.L.; Cocking, R.R., Eds.; National Academy Press: Washington, DC, 2000.
37. *How Students Learn: Science in the Classroom*; Donovan, M.S.; Bransford, J.D., Eds.; National Academy Press: Washington, DC, 2005.
38. Wiggins, G.; McTighe, J. *Understanding by Design*; ASCD: Alexandria, VA, 1998.
39. Fink, L.D. *Creating Significant Learning Experiences*; Jossey-Bass: San Francisco, CA, 2003.

40. Boyer, E.L. *Scholarship Reconsidered: Priorities of the Professoriate*; Jossey-Bass: San Francisco, CA, 1990.
41. Bass, R. *Inventio* **1999**, *1*, 1-9.
42. *Opening Lines: Approaches to the Scholarship of Teaching and Learning*; Hutchings, P. Ed.; Carnegie Publications: Menlo Park, CA, 2000.
43. Cross, K.P.; Steadman, M.H. *Classroom Research: Implementing the Scholarship of Teaching and Learning*; Jossey-Bass: San Francisco, CA, 1996.

Chapter 18

Active Learning in the Introductory Graduate Student Analytical Chemistry Course: Getting Students to "Think Analytically"

Alexander Scheeline and Ryan C. Bailey

Department of Chemistry, University of Illinois at Urbana-Champaign, 600 South Mathews Avenue, Urbana, IL 61801

Active learning benefits graduate students for many of the same reasons as undergraduates. They gain ownership of material from internal motivation, not faculty pull. We describe a first-semester graduate survey course on analytical chemistry that gives students perspective on a wide range of technology, professional ethos, literature awareness, and discusses how to approach novel or difficult topics.

Introduction: Course Goals

From shortly after the Second World War, entering analytical chemistry graduate students at the University of Illinois at Urbana-Champaign (UIUC) took a lecture/laboratory course in Spectrochemical Methods of Analysis in their first semester (*1,2*). This was widely regarded as "boot camp," an intensive, challenging course that taught critical thinking, laboratory technique, measurement concepts, and the specifics of optical spectroscopy. In the late 1980's, it became clear that entering students were typically unprepared for the theoretical rigor of Spectrochemical Methods, and in some cases had so little background in measurement science that an overview course would better serve student needs. Prof. Timothy Nieman thus originated Advanced Analytical Chemistry. In broad brush, this course covered both Instrumental Analysis and Quantitative Analysis at deeper levels than the typical undergraduate offerings, and moved at a breakneck pace. Several faculty taught the course over a 15 year period. One tried using articles from current literature and was astonished at how little ability entering graduate students had in reading primary sources. In 1999, the Department terminated the use of cumulative exams, previously the means by which new students were encouraged to read the literature. An additional change over the decades was the evolution of Analytical Chemistry from a technique-focused field to a problem-focused field. There are now so many tools available (textbook bloat is a convenient indicator of this capability explosion (*3-5*)) that it is impossible to cover them all in one semester.

When one of the authors of this chapter faced teaching Advanced Analytical Chemistry in the fall of 2003, ideas concerning active learning had been in sufficient circulation (*6,7*) that it seemed appropriate not only to adapt course content to the changed clientele, departmental requirements, and scientific viewpoints of the time, but also to change pedagogy to encourage students to feel a sense of ownership regardless of their technical, demographic, or coursework background (*8*). This paper, and an earlier exposition (*9*), report our experience in enlisting the first-year graduate students to teach most (though not all) of the "lecture material" in Advanced Analytical Chemistry.

Teaching the Course

The First Three Years

At the beginning of the course, three review or overview articles on widely disparate topics in measurement science were selected to seed discussion. Those used for the first three years are listed in Table I. These papers had several characteristics. First, they were current, having been published after the end of the previous rendition of the course or covering topics not covered in prior years. Second, they were diverse in the problems they addressed, representing a mix of materials/surface problems, biological problems, and either process analysis or some "hot area" such as terahertz spectroscopy. The first two papers used many analytical tools, as is common in solving many analytical problems. This was not the case with the third paper.

The first day of class, the students were asked to read the first paper and be prepared to discuss it during the second period, were advised that they'd be doing much of the lecturing plus writing a term paper. Little guidance was given

Table I. Papers Used to Seed Discussion

Year	*Papers*
2003	Comazzi, S.; Paltrinieri, S.; Spagnolo, V.; Sartorelli, P. "Some Aspects of Erythrocyte Metabolism in Insulin-treated Diabetic Dogs," *Res. Vet. Sci.* **2002,** *72*, 23-27. Tracy, B. "Materials Analysis and Process Monitoring in Megafabs," *Proc. 28th Int. Symp. Testing Fail. Anal.* **2002,** 69-75. Sinfelt, J. H. "Role of Surface Science in Catalysis," *Surface Science*, **2002,** *500*, 923-946.
2004	Beaucage, G.; Kammler, H. K.; Mueller, R.; Strobel, R.; Agashe, N.; Pratsinis, S. E.; and Narayanan, T. "Probing the Dynamics of Nanoparticle Growth in a Flame Using Synchrotron Radiation," *Nature Materials*, **2004,** *3*, 370-374. Aller, M. A.; Arias, J. L.; Nava, M. P.; Arias, J. "Posttraumatic Inflammation Is a Complex Response Based on the Pathological Expression of the Nervous, Immune, and Endocrine Functional Systems," *Exp. Biol. Med.* **2004,** *229*, 170-181. Vazquez-Duhalt, R.; Torres, E.; Valderrama, B.; Le Borgne, S. "Will Biochemical Catalysis Impact the Petroleum Refining Industry?" *Energy & Fuels*, **2002,** *16*, 1239-1250.
2005	Lucas, L. H.; Larive, C. K.; Wilkinson, P. S.; Huhn, S. "Progress Toward Automated Metabolic Profiling of Human Serum: Comparison of CPMG and Gradient-filtered NMR Analytical Methods," *J. Pharm. Biomed. Anal.* **2005,** *39*, 156-163. Pomfret, B.; Stotlz, C.; Varughese, B.; and Walker, R. A. "Structural and Compositional Characterization of Yttria-Stabilized Zirconia: Evidence of Surface-Stabilized, Low Valence Metal Species," *Anal. Chem.* **2005,** *77*, 1791-1795. Kammer, F. V. D.; Baborowski, M.; Friese, K. "Field-flow Fractionation Coupled to Multi-angle Laser Light Scattering Detectors: Applicability and Analytlical Benefits for the Analysis of Environmental Colloids," *Anal. Chim. Acta*, **2005,** *552*, 166-174.

on what to look for in the papers. Since one frequently does not know the point of a paper when finding one at random in the literature, figuring out its purpose is one part of learning to read. A survey was taken concerning students' background in chemistry, computing, and public speaking. The remaining class time was devoted to a lecture on the history of analytical chemistry, giving a timeline of the development of the field from Lavoisier to automated instrumentation. The change in the field's focus from wet chemistry (through the early twentieth century) to instrumentation (post-World War II through about 1985) to problem solving using the large number of existing tools (recent times) was highlighted.

The second lecture opened with the admonition, "I'll read through the paper, and when I get to a word, sentence, idea, or phrase you don't understand or that you think deserves comment, stop me." One year, they earned a point for stopping the reading; other years they did not. Grading this exercise doesn't seem to be important. What is important is that reading should keep going until EVERY hand has been raised. This took between 2 sentences and 3 paragraphs, depending on the article and group. Inevitably, every student had had to admit in front of the instructor and class that they couldn't get through the first page, much less the entire article. This leveled the playing field. No one could maintain the myth that they were smarter than the rest of the class, and no one had to feel badly that they couldn't understand the article. The rest of the period was spent in explaining how to use the on-line chemistry library, how to dissect sentences and paragraphs to identify words and concepts, and then how to link the two to clarify obscure ideas. Each student was assigned to give a 5 minute presentation on one of the words or phrases that had blocked the class. As instructor, it was important to put the topics in a logical order. Students were matched to the topics list in alphabetical order. Random draw would have accomplished the same goal – forcing the typical student to confront a narrow, but unfamiliar topic not of his or her choosing. Initially, students were given the choice of presentation format. Chalk talks rapidly gave way to PowerPoint presentations. Student pressure led to posting the presentations on-line in advance of the class. Peer pressure accomplished what faculty harangue rarely does – student lectures were typically ready on time without excuses. Only after the course had been taught for several years did we learn that this approach is known as reciprocal teaching (*10,11*).

The initial five-minute lecture was timed, with a point deducted if the student ran over the time limit. Grading was by the instructor (years 1-3) and jointly by the class and instructor (year 4). Content, grammar, spelling (on PowerPoint slides), pronunciation, distracting habits, and slide design were critiqued. Such general feedback led to ever-clearer presentations. One hope was that course alumni would be better prepared to give the second-year literature seminar than had been their predecessors. This hope was not realized. Bad habits remained, as if what was learned about presentations applied only to the Advanced Analytical Chemistry course, not to presentations in general.

Initially, students were told they could discuss their presentations before posting them, but that pre-viewing was optional. Less than ¼ of the class took advantage of this open door policy. At the end of the first year of the course, student feedback indicated that required pre-viewing, especially for the first lecture, reduced anxiety. In subsequent years, the 5 minute talks were all reviewed before posting, and many later lectures were also appraised by the instructor before posting. On occasion, this led to greater rigor. Mostly, instructor vetting of presentations simply reassured the students that they were working in a worthwhile direction. Mathematical detail and rigor were notable by their absence from first drafts of lectures. Even with faculty encouragement, intensive mathematics rarely appeared unless inclusion of chemometrics or differential equations was unavoidable. Perhaps the remark of a former colleague that "Analytical is P. Chem. without the equations" (*12*) is a more general perception than one would hope.

Lecturers were expected to provide a list of questions remaining from or inspired by their presentations; a small part of the grade was based on the quality of these questions. By organizing these questions, topics for 20 minute lectures were established. Students were allowed a minimum of one week to develop these lectures, so in the interim, the course instructor lectured on topics that were important for materials characterization or biological characterization, but that would not be covered in discussing the review article at hand. Typically, two rounds of 20 minute lectures were required to cover all the material in the initial review article. Thus, after 40% of the course was complete, each student had presented 3 lectures and (amazingly enough) covered much of the material that a regular lecture course would cover. At least for the topics of the lectures they gave, students understood the material well. While the number of prepared lectures given by the instructor was small, almost every talk led to a spontaneous mini-lecture to amplify on what the students had said. This approach to teaching is not for those who think slowly on their feet. For those who love to expound spontaneously, it is tremendous fun. Such minilectures were typically 2-3 minutes in length, targeted at issues that would be in-context on the spur of the moment, and, when possible, pointed students to specific authors or literature articles for further details. Towards the end of the first rendition of the course, a student commented, "it seems like there are certain papers that you know off the top of your head, even though you don't do work in mass spectrometry or NMR. Could you give us a list of those core papers?" Such a list was provided. The list was distributed early in the course in succeeding years. When students had additions to make, based on their undergraduate research and classroom experience, the list was extended. It was noteworthy that a student added the Marquardt non-linear fitting paper (*13*) to the "greatest hits" list. It has twice as many "hits" in the Science Citation Index (*14*) as does the Savitsky-Golay algorithm that made the original list (*15*).

When the students began reading the second paper, they found that they were just as clueless about the content as they had been with the first paper. They were not, however, as intimidated. They quickly identified potential topics and went to work generating lectures. In 2005, discussion of the second paper took so long that there was no time to cover the third paper. In the other years, brief coverage of the third paper proved feasible.

In addition to the lectures, students wrote a term paper. They were instructed to find a real societal problem (i.e. a problem that a non-scientist would recognize as a topic where systematic, rational understanding would improve quality of life), identify an aspect of that problem for which measurement science was an important component of the solution, and then focus on how the specifics of the measurements led to a valid solution. Occasionally, a student would try to the lazy path of building a paper around their undergraduate research experience. This was easily nipped in the bud when reviewing proposed paper topics. Topics were due one month into the course, outlines after two months, and rough drafts after 2.5 months. The drafts were diligently edited by the instructor. Content, focus, grammar, appearance, reference quality and quantity were all marked up. Grading was lenient. While a few papers were excellent at this stage, most had severe problems. One student's previously unidentified dyslexia was flagged at this stage; the student was referred to counseling and gratefully reported several months later that a life-long problem was well on the mend. It was at this stage that a surprisingly large fraction of east-Asian students were "caught plagiarizing." The standards for plagiarism are culturally based (*16*), and many areas of the world look at transcription of others' words differently than does the U.S. academy. Most students, when apprised that they'd broken the rules, were shocked, terrified, and intensely stressed. When it was explained that the zero grade for the draft wouldn't sink their careers provided they learned from experience, there was tremendous relief and, in most cases, conformance to western standards of citation thereafter. In one case, the lesson was not learned. Interestingly, that student chose to leave the program after one year in residence.

The final version of the paper was reviewed by three other students. A list of criteria for critiquing manuscripts, listed in Table II, was provided to help focus comments. The students' reviews included a grade that counted towards the reviewed paper's total score. In addition, the instructor graded the reviews. The latter ensured that the reviews would be neither white-wash nor hypercritical; the former meant that the reviewing had impact beyond the score an individual obtained. Some students suggested that more reviewing, earlier in the course, would be desirable. Such reviewing is now a part of the course under the junior author's guidance. This approach has much in common with calibrated peer review (*17,18*). The chief difference is that the students do not see each others' papers until the roughest edges have been worn off by instructor feedback. It was felt this would shelter students with the poorest writing ability from highly negative peer opinion (provided they took the instructor's advice). We have no controls or student polling data to determine if such shelter was noticed or valued.

Table II. Term Paper Review Criteria

1. Does the topic matter? Does the author tell you why it matters? Is the reasoning sound? "This paper fills a much needed hole in the literature." H. D. Drickamer
2. Is the literature background adequate? Are the appropriate authorities cited? Is the focus correct, considering the rest of the article? Citing review articles vs. primary literature.
3. Is the approach to solving the problem appropriate? What might be better? If not the best approach, is it an approach that at least gives perspective not otherwise obtainable?
4. Are procedures described in adequate detail?
5. Are Figures appropriate (*Anal. Chem.* bans straight-line graphs)? What should go in the article and what in supplementary material?
6. Are the statistics adequate, both choice and use? Is the conclusion statistically valid? Are artifacts adequately considered?
7. Are conclusions connected to the data? Has sufficient data analysis been done? Too much? Qualitative vs. quantitative outcome.
8. Is the venue correct? I.e. does it belong somewhere else?
9. Check for mechanics: grammar, spelling, terminology.
10. When in doubt about novelty, do your own literature check.
11. Don't forget a "sanity check."

The first year, the only examination was the final exam. Students felt this was unfair, as they couldn't anticipate the instructor's question style. The second and third year, there was an hour exam focused on the first review article (and the attendant student lectures) as well as a final (comprehensive). In all cases, students were given articles from recent issues of *Analytical Chemistry* or *Analytica Chimica Acta* to read in advance of the exam. They were then asked questions on the new articles as well as on lecture material. They were allowed to talk with anyone in any detail they liked prior to the exam, but had to independently critique the papers during the exam. We were thus able to test if they had learned to read the literature. In a nod to common professional practice, a copy of the *Handbook of Chemistry and Physics* (*19*) was available during examinations. Because many students did not have laptop computers, exams were answered freehand on paper. Taking examinations electronically might be feasible in the future. Lack of ability to block student-to-student communications over wireless connections gives one pause in carrying through such a transition.

One particular final exam question was answered in a most gratifying way. A paper from *Analytical Chemistry* was presented for critique. The students clearly understood what the authors were trying to do, but they savaged the paper. While no two students had the same set of complaints, they were uniform in pointing out omissions, contradictions, and ambiguities. It is unfortunate that the students weren't reviewers of this paper. It was evident to the instructor that the author must have known the answer to every question the students raised, yet every error identified by the students was real. Had the real-life reviewers been as careful as the students, the authors could have spent very little time, yet made the paper more understandable. At least for that group of students, reviewing is now a demonstrated skill.

In 25 years at the University of Illinois, the senior author of this paper was named to the "Incomplete List of Faculty Whose Teaching is Rated as Excellent" only once – following the third iteration of the active learning version of Advanced Analytical Chemistry. Based on average grades on final exams, the students learned best when they did most of the work themselves. However, the style of exam was so different for the two course styles, a strictly numerical comparison is likely misleading.

Year 4: Building on a Good Idea

Upon accepting a faculty appointment at UIUC and accepting a teaching assignment, the junior author of this paper began to think about how to teach graduate level Analytical Chemistry. Immediately turning to my colleague, the senior author, the success of the "active learning" format seemed an attractive alternative compared to more traditional instructor-led forced textbook immersion methods. This structure was also much more comfortable, as it was much more in-line with this author's personal learning style. Accordingly, this general course format was adopted.

The course was structured similarly to that described above, with student presentations driving topic coverage throughout the semester. Students were instructed to pre-read the first assigned paper(s) and come in with a list of unknowns – concepts which they weren't already familiar with – a practice that preceded each round of presentations. Following each "selection session" the class compiled an impressive list of topics, however the process changed dramatically over the course of the semester. The first couple of sessions required lots of prodding by the instructor, while subsequent sessions required lots of quick writing. Upon turning around from the board, every single hand was enthusiastically raised.

One notable difference in this version of the course was the manner in which the topics for student presentations were selected. In order to relate topics covered in the course to the cutting edge research, topics for the first three rounds of presentations were taken from recently published manuscripts authored by external speakers scheduled for the current semester's Analytical Chemistry seminar series at UIUC. Secondary objectives of this selection method were to help students obtain a better understanding of presentations given by invited faculty from other institutions and to remove some of the awkwardness from traditional "meet the speaker" sessions.

The other main difference was the inclusion of student critiques as a component of assigned grades. The students comprising the audience were asked to grade each presentation based upon the criteria of content, quality of presentation and usage/diversity of literature references. Selected student comments were then included on a grading sheet returned to each presenter and accounted for 10% of the assigned grade.

Each student gave three speaker-based presentations throughout the semester; one five minute and two ten minute talks. A final presentation was also included in which students were instructed to give a technique-centric overview of an assigned paper from recent issues of *Analytical Chemistry*. As of the time of submission of this manuscript, the course is still in progress, however, preliminarily, it appears as though the quality of student presentations has improved, for the most part, throughout this experience, in some part due to comments made in student critiques. Students seemed to really care about what their peers said and made efforts to correct these issues – perhaps focusing more on these than instructor comments! Furthermore, the "quality" of critiques seemed to increase dramatically.

All in all, while not completed at as yet, this version of "active learning" graduate analytical chemistry appears to have been a success. Informal surveys, admittedly biased by instructor inquisition, have provided significant positive feedback. Students appear to have appreciated the opportunity to improve their oral presentation skills – especially in light of their second year literature seminars – and seem to feel more confident with speaking publicly about science. Additionally, students seem to have a greater appreciation for the external seminar speakers and have, on occasion, related disparate course topics back to these presentations.

Observations and Suggestions

By having students identify what they find difficult in reading the literature, class time is focused on what the students need to learn (as self-reported), rather than on what the instructor chooses to discuss (which may focus on what students already know; one finds out whether the focus was right only in retrospect). The students ensure that little time is wasted. This alone tunes the class to its instantaneous audience.

Faculty knowledge breadth is sorely tested in this environment. Students may overlook areas that faculty can quickly remedy, but they may also completely misconstrue a paper, something that only an experienced faculty member can recognize and fix "on the fly." Previewing lectures may help spot trouble areas where the faculty needs to be prepared for discussion, but it is easy to be blindsided by theoretical or practical holes in student understanding. While there is little problem with saying, "I don't think that's right; I'll revisit this at the beginning of the next class," one must avoid letting misconceptions slide, only to revisit an issue weeks later after the misconceptions have solidified in student minds. We instinctively know this in preparing "normal" lectures. Here, the challenge is to anticipate the wide range of student viewpoints so that, at random, a 2 minute clarification on any discussion item is available. Fortunately, no one expects a polished PowerPoint presentation in spontaneously answering a question. Preparation time is spent on content, not computer file generation. Blackboards or white boards have been the presentation tools, though one could imagine using a tablet PC.

Timeliness of material is important. In selecting focus papers, only material that had appeared since the end of the previous year's course was considered. This ensured that whatever backfiles previous students had kept would be of no use in succeeding years. It also gave the material an immediacy that some fixed set of articles would lack.

References Cited

1. Crouch, S. R.; Enke, C. G.; Walters, J. P.; Hieftje, G. M. *Appl. Spectrosc.* **2004**, *58*, 165A-172A.
2. Horlick, G. *Spectrochim. Acta* **2006**, *61B*, 602-618.
3. Robinson, J. W. *Undergraduate Instrumental Analysis*; 2 ed.; Marcel Dekker: New York, 1972.
4. Robinson, J. W. *Undergraduate Instrumental Analysis*; 4 ed.; Marcel Dekker: New York, 1987.
5. Robinson, J. W.; Skelly Frame, E. M.; Frame II, G. M. *Undergraduate Instrumental Analysis*; 6 ed.; Marcel Dekker: New York, 2005.

6. Pinkava, A. *Nursing Outlook* **1985**, *33*, 257-258.
7. Ross, M. R.; Fulton, R. B. *J. Chem. Educ.* **1994**, *71*, 141-143.
8. Walters, J. P. *Anal. Chem.* **1991**, *63*, 977A-985A.
9. Scheeline, A.; *Active Learning in Advanced Analytical Chemistry, a first year course for graduate students*; Analytical Sciences Digital Library, www.asdlib.org, #10039, **2005**.
10. Forrest, S. *J. Cont. Ed. Nursing* **2004**, *35*, 74-79.
11. Krych, A. J.; March, C. N.; Bryan, R. E.; Peake, B. J.; Pawlina, W.; Carmichael, S. W. *Clin. Anatomy* **2005**, *18*, 296-301.
12. Washburn, N.: Madison, WI, 1978, personal communication.
13. Marquardt, D. *SIAM J. Appl. Math.* **1963**, *11*, 431-441.
14. *Science Citation Index*, Institute for Scientific Information, accessed December 7, 2006.
15. Savitsky, A.; Golay, M. J. E. *Anal. Chem.* **1964**, *36*, 1627-1639.
16. Wang, Y.; Martin, M. A.; Martin, S. H. *College Teaching* **2002**, *50*, 97-101.
17. Russell, A. A.; Chapman, O. L. *221st ACS National Meeting*; American Chemical Society: San Diego, CA, 2001; CHED-009.
18. Russell, A. A.; Chapman, O. L. *222nd ACS National Meeting*; American Chemical Society: Chicago, IL, 2001; CHED-166.
19. Lide, D. R. *Handbook of Chemistry and Physics*; 73 ed.; CRC Press: Boca Raton, FL, 1992.

Chapter 19

Collaborative Research: The Good, the Bad, and the Beautiful

William R. Heineman[1] and Richard N. Zare[2]

[1]Department of Chemistry, University of Cincinnati, Cincinnati, OH 45221–0172
[2]Department of Chemistry, Stanford University, Stanford, CA 94305–5080

Interdisciplinary research is on the rise, driven by the lure of solving increasingly complex research problems and a shift in funding trends. The resulting need for broader technical expertise than typically exists in a single academic research group has stimulated collaborative projects, which means they are becoming the vehicle for teaching graduate students how to do research. Here we describe personal observations about the impact of collaborative research on faculty, postdocs, and graduate students.

Introduction

Knowledge is available as a whole. Universities by their composition disintegrate knowledge department by department. Interdisciplinary efforts introduce an important change in mindset to reintegrate information for the purposes of discovery. Increasingly, researchers are becoming aware that solutions to problems do not come with names of departments written on them, but rather require contributions from experts from different fields (*1*). Groups that work in interdisciplinary research are able to multiply their efforts and thus have the power to solve many more problems. Yet, as this chapter shows, major impediments exist to these efforts. Here we describe personal observations about collaborations, in particular, the impact of the collaborative mode of research on faculty, postdocs, and graduate students.

Although the success of any interdisciplinary effort requires that departments look outward instead of inward, we are not advocating weakening departments to strengthen interdisciplinary efforts. Indeed, to be successful, these efforts further require that departments be independently capable.

To begin, we examine the drivers for the increased emphasis on collaborative research. First, federal agencies have made available more funds for this type of research. As Federal agencies turn their perspective from long-term fundamental advances to short-term applications, the focus of attention naturally shifts from simple to complex problems. Simple problems are readily assigned to specific departments that serve as important keepers of the truth for their disciplines. Complex problems by their nature demand input from multiple experts who often reside in different departments. Second, there is a growing realization that some of the most exciting problems lie at the boundaries of traditional fields. Third, the dramatic increase in knowledge and sophistication (complexity) of experimental techniques has led to increased specialization. Thus bringing the full powers of all applicable techniques to bear on a problem can require assembling a team of appropriate experts. Fourth, researchers are encouraged to launch interdisciplinary programs because industry states that it wants graduate students who are trained to work on teams as problem solvers. Finally, what the information technology revolution has done to make collaborations possible cannot be underestimated. Once upon a time, researchers exchanged information by mail (in the form of lengthy letters). Telephone connections were possible but prohibitively expensive. Today, communication throughout the world is almost instantaneous by fax, e-mail and even video conferencing at almost no cost. An NMR obtained in one country can be analyzed in another country on the same day. Even experiments can be controlled remotely.

Recent reorganization of the Engineering Directorate at NSF (ENG) clearly illustrates this trend toward collaboration (*2*). ENG is in the final stages of a comprehensive reorganization in which three crosscutting units are being created to facilitate multidisciplinary research. The Office of Emerging Frontiers in Research and Innovation (EFRI) is a major new element that will fund those research opportunities that were difficult to fund under the pre-existing structure. EFRI will focus on "Interdisciplinary opportunities with high potential payoff leading to new research directions and areas that result in a leadership position and/or significant progress on a recognized national need or grand challenge" (*2*). The changes in funding in ENG are illustrated in Figure 1, which shows that the increase in funding awarded to collaborations has been at the expense of single investigator grants. As stated in 2001 by then Acting Director of the National Institutes of Health Ruth Kirchstein, "The kind of science projects we and the whole world do now...are not small laboratory projects between an investigator, one technician, and two graduate students. They are projects that typically need a lot of resources." (*3*).

Of course, collaboration in science is not a new idea. It is especially common in industrial laboratories where assembling teams to work on projects has been routine for some time. However, collaboration has been far less common in academics. For a long time, there was the ivory tower model: a single professor surrounded by postdocs and graduate students who learned from the professor. At some point, a committee materialized and then approved a Ph.D. dissertation for a graduate student. This is not to say that collaborations have not existed in academics. The Watson and Crick collaboration to solve the structure of DNA is one well-known example. But, in general, collaboration has been far less common in academics than in industry. Why?

The research culture and value system in academics are different from those in industry. Team players who contribute to company profitability are highly valued in industry. Managers can "order" a collaboration---a highly effective approach to "getting things done". However, in the academic world, professors (at any rank) are basically independent entrepreneurs (rugged individualists) who succeed or fail largely on the basis of their individual efforts to establish a strong research program as judged by their peers. Here the reward system comprises novelty of ideas, significance of publications, and prizes, especially the Nobel. Individual credit is the currency of promotion and reputation. The notion of a professor being ordered to work on a project is foreign to this culture. Thus, collaboration is not as easy and has actually been regarded as risky for a young, unestablished investigator, mostly because of perceived vagueries in defining their individual contribution to the field. But the winds of change are blowing. What are the implications for faculty and their students?

A Successful Collaboration?

We like the collaboration shown in Figure 2.

The cartoon shows a diverse team comprising individuals with a history of noncooperation. However, they have a clear, shared goal whose importance

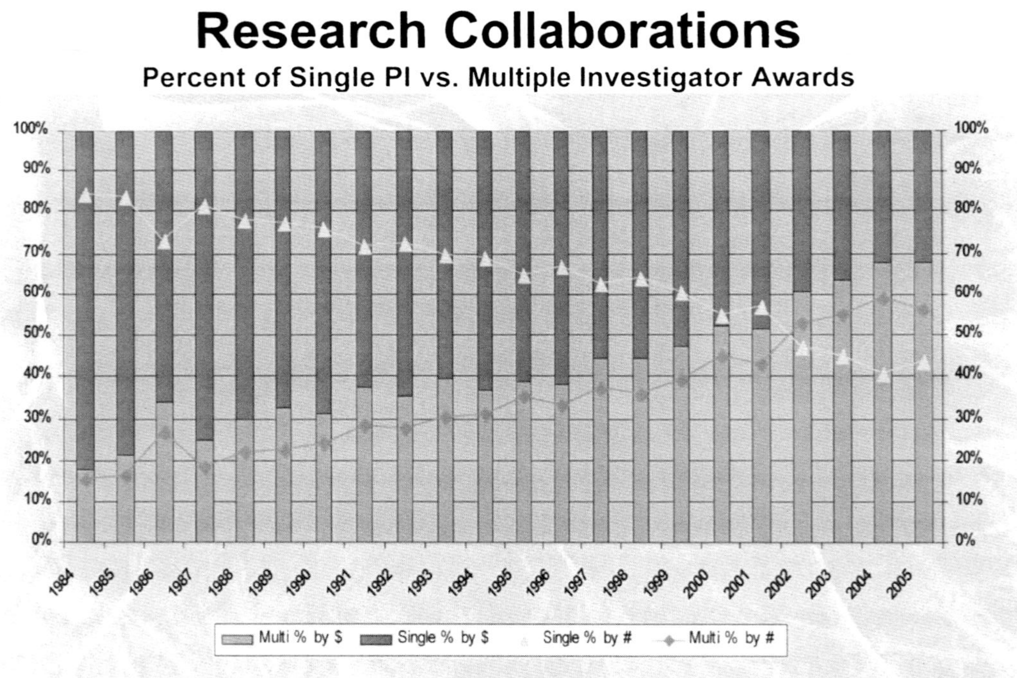

Figure 1. Trends in single investigator vs. multiple investigator awards. (Adapted with permission from reference 2. Copyright 2006.) (See page 2 of color inserts.)

Figure 2. Cartoon depicting elements of a successful collaboration. (Adapted with permission from reference 4. Copyright 2006.) (See page 2 of color inserts.)

surpasses their individual differences. Having such a goal provides the glue that can hold a team of diverse collaborators together. As shown in the cartoon, a good plan for achieving success is also in place, demonstrating another vital component to a successful collaboration. This team is about to succeed in getting beer out of the refrigerator by working together. But do they have a plan for opening the beverage containers? A real test of the collaboration will determine whether they have all succeeded. Will the larger collaborator decide the collaboration is no longer needed and drink all the beer (he seems to be thinking about that now). Or will he look ahead to the next time they can all work together? Depending on his decision, this could be a successful long-term collaboration with all parties satisfied or a one-time collaboration with only one party satisfied.

In our experience, collaborations work best when one or more of the following elements are in place:

- *Clearly defined goal of high importance.* A clear goal that all parties have a high interest in achieving must be defined. All parties must make a creative contribution that is important to their own fields. The project cannot be on the "back burner" for any team member.

- *Members working together.* The goal cannot be accomplished individually. Working together will enable all members to multiply their individual accomplishments.
- *Confidence in technical abilities.* Team members must respect each other's abilities in their respective fields. Members must cover their own areas.
- *$$$$.* A large grant can be a powerful motivator. Of course, the team will want to renew the grant, which can serve to keep the effort on track.
- *Suitable personalities.* Suitable personalities are important to long-term collaborations. The ability to work together keeps the team together. One author observed a collaboration that lasted only two days---until the first disagreement between two strong personalities.

The existence of these elements in a collaboration greases the skids of success.

Impact on Faculty

Although collaborative projects have the significant advantages discussed above, becoming involved in a collaboration has definite implications for faculty.

Receiving proper credit for one's contributions is perhaps the most important element because of the academic culture and value system. Assigning credit is relatively easy in a project involving a single professor and the professor's students. However, as other professors/scientists become involved, assigning credit can become increasingly difficult, especially to the outside observer. Although we all want to be recognized for our contributions, the reasons the contributions of each collaborator must be clearly defined and apparent extend beyond this natural human trait. Assistant professors need to create a clear reputation to justify promotion. Important are the external letters of recommendation in which a reviewer is asked to evaluate a person's contributions to a field. Multiauthored papers present difficulties for the external reviewer who might not know the "inside story" regarding the collaboration. Associate professors also face these same issues when being considered for their next promotion. Success with research proposals, invitations to speak, and the receipt of awards are all based on the recognition of individual contributions, and this applies to professors of all ranks.

Based on our observations, some suggestions for dealing with the issue of proper credit assignment are provided:

- Clearly identifying the contribution of each individual generally becomes more difficult as the technical areas merge, making collaboration within the same department more problematic. Therefore collaborating outside one's own department may be simpler in this regard.
- Ideas are perhaps the most important currency in academics and pedigree should be identified. At the outset, agree who had the original idea upon which the collaboration is based. Clearly define the role and expectations of each collaborator. Having each collaborator make a creative contribution in their own field is the optimum situation. Enthusiasm wanes when parties feel they are merely a "service" in support of another person's ideas.
- Multiple publications allow diverse team members to each publish in "their" journal and provide various opportunities for first authorship for the students/postdocs involved.
- When collaborating with very junior faculty, senior faculty must be sensitive to the junior person's special situation and what is best for career advancement. It is easy for senior professors to get all the credit because of their established reputations. Senior professors must go out of their way to ensure this does not happen.

"There are two ways to do something---the right way and my collaborator's way!" Long-term collaboration is much more enjoyable when the investigators are getting along. Flexibility becomes important. Collaborating with the "control freak" can be problematic.

The dynamics of running a research group and mentoring students are definitely affected by the collaborative mode of research. Communication and organization become more important as the manpower and complexity of a project increase with the addition of other research groups. When multiple faculty members are spending equivalent amounts of time advising one student, co-advising provides a mechanism for giving each member credit. If this happens, the student is no longer "just yours" and considering the other faculty member's point of view becomes necessary. Even though the expectations faculty members have of graduate students vary widely, all students should be able to see the big picture and understand how they fit into it. Each student should be given clear charge of a piece of the big project with dissertation rights and first authorship on papers.

Impact on Graduate Students and Postdocs

We now examine the impact collaborative research has on coworkers in the same research group, both graduate students and postdocs. We address primarily the effect on graduate students, but we believe that the same considerations readily transfer to postdocs. Collaborations have both advantages and disadvantages from the student's perspective. In our opinion, the advantages are considerable. Below are a few.

The Beautiful

The excitement of being associated with a well-organized collaboration that is addressing an exceptionally important scientific problem is the most significant benefit for some. Rubbing shoulders with top people from different disciplines can be a stimulating experience. Seeing the inner workings of a good collaboration provides valuable experience for students intending to follow the collaborative path in their own careers. Inevitably, various points of view emerge about what is significant in tackling the same problem, and these different outlooks and procedures broaden the perspectives and sharpen the problem-solving skills of all those who work together.

The Good

Collaborations can provide considerably broader technical exposure during graduate school. A collaboration often exists because a project requires technical expertise beyond that found in a particular research group. Most collaborations automatically expose the group members to "external" expertise. An example of this exposure is the group meetings where the entire team presents and discusses progress and problems. Such meetings can provide an excellent opportunity for students to broaden their experience by asking questions. Another example is when students from different areas work together and perform joint experiments. This provides real "hands on" experience in the other area. Often students begin to attend seminars in the other departments with which they have collaborated. Of course, the benefits that each student gains depend on the effort expended, which ranges from significant to only the minimum needed to get by.

Greater breadth is good for students to acquire but not at the expense of depth in one's own area. Reaping the full benefits of a collaboration requires more effort by a student. However, this can prove useful later. "Success in industry requires continually developing new skills and contributing in broad areas. Having experienced this in graduate school provides a head start (*5*)." In academics, breadth can be the source of new ideas for young professors to develop their own research programs.

Collaborations provide the opportunity to develop communication and interaction skills, which can be very important to students later in their careers. Communication is key in industrial projects. Effective communication with people from different technical backgrounds is important in industry. The abilities to understand both the entire project and your role in the project, and to explain your work to those who are not expert in your area are critical skills to be learned in graduate school. Collaborations provide students with experience working in a team environment with people from various technical backgrounds and with different problem-solving approaches. "Those who master these skills do the best in industry---accomplish more, are valued more, achieve more personal goals. Already having experience working in a team environment is invaluable for someone entering industry (*5*)."

Although students interact with students from other groups, the extent of these interactions varies greatly depending on the project. For example, in one of our projects, a chemistry student traveled to another university on several occasions for one or two weeks at a time to conduct experiments with an engineering student. The engineering student had developed a device, whereas the chemistry student had developed some biological assays. The goal was to combine the engineer's device with our chemistry and conduct enough experiments to produce a paper. In this case, the engineers could make devices that we could not, and we had biological assays that they did not have. Thus this collaboration exemplified one of the elements for good collaboration: Members each contribute to and control one part of the project within their area of expertise. Because the chemistry student traveled to the other university to do the experiments and wanted to minimize the amount of travel time for cost and personal reasons, the two had to use their time together efficiently. This required considerable planning and coordination of effort. Both students had to have their part of the experiment ready. In the case of the chemistry student, she had to ensure she had everything needed because she would be away from her environment where the usual equipment was readily available. Once together, they had to work cooperatively. Even though the students were not only from different technical backgrounds, but also from different countries, the collaboration worked extremely well. Once the experiments were started some unanticipated problems were encountered, but the students worked together and solved them. Three joint publications resulted from this collaborative project over the course of two years. This collaboration is one example where the efforts of the collaborators are multiplied rather than divided. Several years after graduating and working in industry, the chemistry student commented: "I think back about that collaboration often and am so fortunate to have had the experience during graduate school. As challenging as it was, I learned so much about communication across cultures and disciplines, a skill that is critical to being productive in industry (*6*)."

The Bad???

Here are some problems to watch out for. The responses are direct quotations from former students who have had successful careers in industry/government laboratories.

They are holding me back! A student's progress can be hindered by the performance of other group members. Friction can develop when one student is relying on another student who is not producing at the expected level. This can happen when students are mismatched in technical expertise, motivation, or work ethic. "True, but this is also what you find in industry. Your efforts and work are often dependent on others---team members, managers, those who report to you. Almost everything is interlinked and interdependent in industry (*5*)."

How do I fit into the big picture? As stated above, the contributions of all collaborators must be clearly defined and apparent to them. This applies to students as well as professors. Graduate students need clearly defined projects that they can call their own. (What is clearly defined in the faculty advisor's mind is not necessarily so in the student's!) Students need to know that this material is for their own dissertation. Ownership truly matters in having a successful outcome to collaborative projects.

I should be first author on this paper--not one of them! There can be only one first author on a paper! A graduate student or postdoc is typically the first author on a scientific publication. In coauthored papers with equally important contributions from two or more students/postdocs, first authorship can be a sore point of contention. Producing multiple papers from the students' work solves this problem somewhat by making rotation, if appropriate, of first authorship possible. Ideally, collaboration would enable students to accomplish more. As a result, they would be able to produce the usual number of papers with first authorship as well as additional papers.

Who's dissertation does this go into? There will be some overlap in dissertation material among the students, but the emphais will be different. For example, engineering students would want some results from a chemistry student in their dissertation to validate their engineering projects. Likewise, chemistry students would want a nice design of the device in their dissertations to illustrate why/how the experiments were done. Giving proper credit to the other person is critical. Nothing collapses a collaboration more rapidly than if one party feels that they are only working for the other with no appreciation for their efforts.

I'm working a lot harder than students on other projects. Generally, but not always, more effort is required to understand a broader array of technical issues/areas. For some, it can be intimidating to work with others whose expertise is greater than their own in an area. "True, but in industry the norm is to work on multifunctional, multidisciplinary teams with varying levels of expertise and opinions (*5*)."

I'm having trouble working with these people. It is more complicated to work across multiple groups with technical, language, cultural, work-ethic, and personal differences. "True, but in reality, it mirrors the business world. Today's work force is diverse (*5*)."

But my coadvisor said just the opposite. Conflicting advice from multiple advisors happens. One student found this "..initially confusing, but ultimately beneficial because it forced me to come to my own conclusions (*6*)."

Meetings, meetings, meetings. "I have learned to value all those group meetings that had nothing to do with my own work. I learned how to focus on very different subject matter. The students may grumble about being stretched and about the number of meetings. They will be glad for it later (*7*)."

Institutional Changes

Thought must be given at the highest level to foster collaborations. This list of recommendations provides a starting place.

- Establish a "culture of collaboration."
- Give due credit to all co-PIs on multi-investigator grants. The restriction of only one PI on a grant is obsolete. All names should be listed in announcements.
- Administer interdisciplinary funding so that each department has its own budget and receives full credit for its share of the grant.
- Give multiple-authored publications equal weight with single-authored papers.
- Review promotion and tenure documents for bias against collaboration.

Conclusions

We advocate collaboration as the way of the future. Graduates entering the industrial sector, where collaboration is already the norm, benefit from working on collaborative projects as students. This benefit will increasingly apply to graduates entering academics as that sector gradually transforms from the historical single-investigator project to more multi-investigator projects.

"Collaborative research shifts the balance on individual mastery to working effectively with others. Both are crucial elements for most successful scientists (*8*)."

Diverse groups working together, making creative contributions to complex problems at the cutting edge, and extending themselves to individually unreachable heights is collaboration at its best.

Acknowledgments

Richard Zare thanks the Howard Hughes Medical Institute for a HHMI Professorship in support of advancing teaching and outreach.

We acknowledge the 2006 Oesper Symposium at the University of Cincinnati honoring Richard Zare for precipitating this collaboration between the authors.

William Heineman acknowledges former students who participated in a questionnaire regarding the impact of collaboration on their careers. Their comments were most helpful!

This paper is based in part on talks presented by William Heineman in symposia at the 2005 National ACS meeting in Washington, D.C.; 2001 National ACS Meeting in Chicago; and 1997 National ACS Meeting in Las Vegas.

References

1. Metzger, N.; Zare, R.N. *Science* **1999**, *283*, 642-643.
2. Senich, D. *Engineering at the National Science Foundation*, IEEE LEOS Meeting, Montreal, Canada, October 31, 2006.
3. Kirschstein, R.L. *Chem Eng News*, August 6, 2001, p. 30.
4. Stone, J.R, III. National Research Center for CTE, University of Minnesota. http://www.nccte.org/webcasts/media/20041122/index_files/TextMostly/index.html (accessed December 6, 2006).
5. Wehmeyer, K. Personal communication, August 1, 2005.
6. Thomas, J. Personal communications, August 17, 2005 and November 30, 2006.
7. Slaterbeck, A. Personal communication, August 9, 2005.
8. Doyle, M. Personal communication, August 7, 2005.

Chapter 20

Final Thoughts

Patricia Ann Mabrouk

Department of Chemistry and Chemical Biology, Northeastern University, 360 Huntington Avenue, Boston, MA 02115

Anticipating your enthusiasm for introducing active learning into your classroom and teaching laboratory, you will find below a short list of suggestions for how to get started as well as a brief bibliography of archival resources on teaching and learning.

If you are considering active learning, a good place to begin is by visiting a class or laboratory section using active learning taught by a colleague.

When you are ready to try your own active learning experiment, start small with one well considered active learning exercise. Don't try and change your whole course all at once.

Plan your exercise and how you will assess the success of your experiment. Consider using the principles of Backward Design to guide you:

1. Begin by identifying what you want your students to know/be able to do (outcomes);
2. Next ask yourself what characteristics you want them to exhibit as a result of acquiring this knowledge or skill? (evidence); and
3. Finally, identify the classroom and/or laboratory activities that will satisfactorily demonstrate that this knowledge or skill has been acquired (assessment).

Discuss your plans with others interested in active learning and teaching in your department, institution, or at nearby institutions. Carry out your exercise and then reflect on your work.

Consider partnering with others in your department or in other departments at your institution who are experts in teaching and learning in order to design and implement meaningful experiences for your students.

If your department currently doesn't use active learning methods and primarily uses lecturing as an instructional method (generally a passive technique), expect your students to be somewhat resistant to active learning as they are likely unfamiliar with it and concerned with how it will impact their performance in your course. Prepare your students by explaining your objectives and the potential benefits to them. Listen carefully to what your students have to say.

If you are considering using team work, don't try to reinvent the wheel. Use the protocols suggested by a local expert who regularly uses group work either in your department or in another department at your institution. See how things go (*i.e.*, assess!) and then try another collaborative exercise and change it, if warranted, based on the feedback you collected. This way you will know that your students are learning and that they are satisfied with the instructional process.

Bibliography

The sources below are good starting points for further study on the topics discussed in this volume:

General Resources on Learning and Active Learning

Barkley, E.; Cross, K. P.; Major, C.H. *Collaborative Learning Techniques: A Handbook for College Faculty*; Jossey-Bass: San Francisco, 2004.

Pienta, N. J.; Cooper, M. M.; Greenbowe, T. J. *Chemists' Guide to Effective Teaching*; Prentice Hall: New York, 2005.

Grasha, A. *Teaching with Style*; Alliance Publishers: Pittsburgh, 1996.

Johnson, D.; Johnson, R.; Smith, K. *Active Learning: Cooperation in the College Classroom*; Interaction Book Company: Edina, 1991.

National Research Council *How People Learn: Brain, Mind, Experience, and School*; National Academy Press: Washington, D.C., 2000.

Cooperative Learning

Kagan, S. *Cooperative Learning;* Resources for Teachers, Inc.: San Juan Capistrano, 1992.

Team-Based Learning. A Transformative Use of Small Groups; Michaelsen, L. K.; Knight, A. B.; Fink, L. D., Eds.; Praeger: Westport, 2002.

Adult Learning Theory

Brookfield, S. D. *Understanding and Facilitating Adult Learning*; Jossey-Bass: San Francisco, 1986.

Knowles, M. S.; Holton, E. F., III; Swanson, R. A. *The Adult Learner*; 5 ed.; Butterworth-Heinemann: Woburn, 1998.

The New Update on Adult Learning Theory; Merriam, S. B., Ed.; Jossey-Bass: San Francisco, 2001; Vol. 89.

Problem-based Learning

Bringing Problem-Based Learning to Higher Education: Theory and Practice; Wilkerson, L.; Gijselaers, W. H., Eds.; Jossey-Bass: San Francisco, 1996; Vol. 68.

The Power of Problem-Based Learning; Duch, B. J.; Groh, S. E.; Allen, D. E., Eds.; Stylus: Sterling, 2001.

Case Study Method

Barnes, L. B.; Christensen, C. R.; Hansen, A. J. *Teaching and the Case Method: Text, Cases, and Readings*; third ed.; Harvard Business School Press: Boston, 1994.

Service Learning

Service-Learning in Higher Education: Concepts and Practices; Jacoby, B., Ed.; Jossey-Bass, Inc.: San Francisco, 1996.

Assessment

Angelo, T. A.; Cross, K. P. *Classroom Assessment Techniques, A Handbook for College Teachers*, 2 ed.; Jossey-Bass Publishers: San Francisco, 1993.

Fink, L. D. Creating Significant Learning Experiences: An Integrated Approach to Designing College Courses; Jossey-Bass: San Francisco, 2003.

Tanner, D. E. Assessing Academic Achievement; Allyn and Bacon: Boston, 2001.

The National Institute for Science Education's (NISE) College Level One (CL-1) Team. The Field-Tested Learning Assessment Guide (FLAG). http://www.flaguide.org/ (accessed December 11, 2006).
Walvoord, B. E. Assessment Clear and Simple; Jossey-Bass: San Francisco, 2004.

Digital Libraries

Analytical Sciences Digital Library. http://www.asdlib.org/ (accessed December 11, 2006).
National Science Foundation. National Science Digital Library. http://nsdl.org/index.php (accessed December 11, 2006).
Multimedia Educational Resource for Learning and Online Teaching. http://www.merlot.org/merlot/index.htm (accessed December 11, 2006).

Peer-reviewed Journals Focused on Science Education Research

Journal of Chemical Education
The Chemical Educator
Journal of College Science Teaching
Science Education

Indexes

Author Index

Subject Index

D

H

I

Q

R

S

T

U

V

W